Mechanical Engineering Series

Series editor

Francis A. Kulacki, University of Minnesota

The Mechanical Engineering Series presents advanced level treatment of topics on the cutting edge of mechanical engineering. Designed for use by students, researchers and practicing engineers, the series presents modern developments in mechanical engineering and its innovative applications in applied mechanics, bioengineering, dynamic systems and control, energy, energy conversion and energy systems, fluid mechanics and fluid machinery, heat and mass transfer, manufacturing science and technology, mechanical design, mechanics of materials, micro- and nano-science technology, thermal physics, tribology, and vibration and acoustics. The series features graduate-level texts, professional books, and research monographs in key engineering science concentrations.

More information about this series at http://www.springer.com/series/1161

Malay K. Das · Partha P. Mukherjee
K. Muralidhar

Modeling Transport Phenomena in Porous Media with Applications

 Springer

Malay K. Das
Department of Mechanical Engineering
Indian Institute of Technology Kanpur
Kanpur, Uttar Pradesh
India

K. Muralidhar
Department of Mechanical Engineering
Indian Institute of Technology Kanpur
Kanpur, Uttar Pradesh
India

Partha P. Mukherjee
School of Mechanical Engineering
Purdue University
West Lafayette, IN
USA

ISSN 0941-5122   ISSN 2192-063X   (electronic)
Mechanical Engineering Series
ISBN 978-3-319-69864-9   ISBN 978-3-319-69866-3   (eBook)
https://doi.org/10.1007/978-3-319-69866-3

Library of Congress Control Number: 2017956310

Printed on acid-free paper

This Springer imprint is published by Springer Nature
The registered company is Springer International Publishing AG
The registered company address is: Gewerbestrasse 11, 6330 Cham, Switzerland

Preface

In recent years, transport phenomena in porous media have attracted significant attention from multiple branches of science and engineering. The reasons for such renewed interest are twofold. First, the past two decades have experienced the emergence of a variety of porous media applications, primarily involving energy, environment, and biological systems. Second, the rapid increase in computing power has facilitated theoretical advancement in porous media transport modeling. In this context, the present monograph appends classical theories of porous media transport with recent advancements in the domain. The monograph aims at providing a strong support to the graduate and senior undergraduate students interested in porous media research.

This book is primarily divided into two parts: theory and applications. Types of porous media and the importance of mathematical modeling of transport through such media are discussed in Chap. 1. While Chaps. 2 and 3 discuss the theories of transport phenomena in porous media, Chaps. 4–7 deal with engineering applications and their modeling considerations. Finally, Chap. 8 summarizes the present understanding as well as the imminent challenges of modeling transport phenomena through porous media.

Modeling of transport through porous media may be undertaken at various scales. Chapter 2 examines the fundamental distinction of continuum- to pore-scale models. The chapter then follows the historical development from the Darcian transport model to the contemporary phenomenological formalisms. Chapter 2 also includes the treatment of multiphase flow and heat transfer through porous media. Finally, charge transport and chemical reactions are also included. In continuation with the continuum models developed in Chaps. 2 and 3 outlines the mesoscale models for transport in porous media. In particular, Chap. 3 delineates the application of Lattice Boltzmann method (LBM) in understanding transport phenomena at the pore scale. The chapter compares relative advantages of LBM in complex geometries, particularly in porous media with the continuum formulation.

The theoretical development, discussed in Chaps. 2 and 3, provides the essential background for the applications contained in Chaps. 4–7. Chapters 4 and 5 describe models of transport phenomena in electrochemical and biological systems. While

Chap. 4 includes the modeling of fuel cells and batteries, Chap. 5 describes blood flow through aneurysms in human arteries. Chapter 6 includes a novel approach of modeling oscillatory flow in heat exchangers within a porous media framework. Finally, Chap. 7 encompasses the reservoir simulation for CO_2 sequestration in hydrate form. Overall, Chaps. 5–7 cover broad areas of contemporary applications in porous media, encouraging interested researchers to delve into their chosen areas.

While this book provides theoretical and modeling support with examples to porous media researchers, it does not intend to substitute well-established monographs on transport phenomena. Fundamental knowledge of fluid mechanics, heat and mass transfer, and electrochemical systems is necessary to efficiently utilize this book. Finally, the references included in this book will provide additional resources of study for early researchers in the subject.

The authors gratefully acknowledge the Curriculum Development for Technical Education (CDTE) cell of Indian Institute of Technology Kanpur for financial support. The authors are grateful to the institute authorities at IIT Kanpur and Texas A&M University for providing a conducive research environment.

Kanpur, India   Malay K. Das
West Lafayette, USA   Partha P. Mukherjee
Kanpur, India   K. Muralidhar
March 2017

Contents

Chapter 1
Introduction

K. Muralidhar

The term *porous medium* is used widely in many contexts and applications. It refers, usually, to an overall solid phase in which a distribution of voids prevails. These voids may be filled with another solid material thus constituting a *composite*. Of interest to the present discussion is a solid phase with fluid-filled voids. The second distinguishing aspect of the present monograph is that it studies *interconnected* void spaces in a solid medium. Thus, fluid flow can take place through the voids, most commonly, along a tortuous path. The study of flow patterns and distribution of other physical variables such as temperature in this structure constitutes the subject of *flow and transport in a porous medium*.

In the development of the subject, certain conditions are enforced so that mathematical simplifications are possible. The first is that the solid material is immobile, thus creating a void space of fixed, time-independent geometry. Forces arise from the flow of a fluid phase through the pores. The second is that the pore scale, namely, the characteristic dimension of the void is much smaller than the overall dimensions of the solid phase across which flow is taking place. In the most general setting, a porous medium will have a primary length scale decided by the application under consideration. In addition, a hierarchy of smaller scales, representing the voids within the solid matrix, may be realized. The pore geometry, however, is taken to be unchanging in time. Effectively, the solid matrix is exposed to fluid forces but is taken to be rigid.

The above simplifications eliminate certain important applications from further analysis. A suspension of solid particles in a liquid medium can give rise to a variety of interesting phenomena, for example, self-organization into networks and clusters. The effective viscosity can be orders of magnitude higher, and scalar diffusivity can be enhanced, jointly introducing time-dependent behavior. Other applications that cannot be addressed using the porous medium approximation include multiphase flow of gas–solid mixtures, steam–water, and gas–liquid combinations, and the dynamics of granular media. Small deformations of a porous medium at timescales much greater than that of the flow field can possibly be

© Springer International Publishing AG 2018
M.K. Das et al., *Modeling Transport Phenomena in Porous Media with Applications*, Mechanical Engineering Series,
https://doi.org/10.1007/978-3-319-69866-3_1

1

accounted in analysis, but this is often an exception. The physical problem thus constructed reduces to flow and transport in a complex geometry.

1.1 Physical Mechanisms

At first glance, fluid flow in a porous medium can be visualized as flow in a three-dimensional tube of varying cross-sectional area. Velocities and pressure in this arrangement can be determined by solving the Navier–Stokes equations, the difficulty in modeling being centered on the geometrical complexity. This step, by itself, is a formidable task. In contexts such as natural reservoirs, the pore scale can be of the order of a fraction of a millimeter while the formation length scale can be over hundreds of meters. Additional difficulties may arise from large-scale features such as faults cutting through the porous region and creating new pathways for flow. It must be said here that the Navier–Stokes approach to porous medium modeling of fluid flow remains far from complete.

The spread of applications in porous media is large and a great many possibilities exist. Each of these will introduce additional mechanisms in terms of forces and fluxes, leading to a revised system of governing equations. Some of these possible situations are listed below.

i. Liquid does not fill the pore space and gas–liquid (or, vapor–liquid) interfaces form. They may further move with time. Interfaces also form when two immiscible liquids are mobilized by an overall external pressure difference.
ii. Part of the fluid is heated and a thermal front moves through the porous medium. The conjugate problem of heat transfer between the fluid and the solid matrix may also be relevant.
iii. Closely related to (ii) is transport of contaminant such as dissolved matter and particulates in a porous medium.
iv. Generalizing (i)–(iii), multiphase, multicomponent flow and transport in a porous medium emerges as a topic of extreme complexity. Chapter 7 of this text describes one such application where phase change is accompanied by chemical reactions.

1.2 Representative Elementary Volume

Modeling flow and transport in the pore space of a porous medium may be considered a first principles approach, the difficulty of which in realistic dimensions and meaningful timescales can be well-imagined. A revised approach that averages flow rates over dimensions much larger than the pore diameter greatly simplifies analysis. Here, a *representative elementary volume* (REV), Fig. 1.1, jointly encompasses the pore space and the solid matrix. It is small in comparison to the

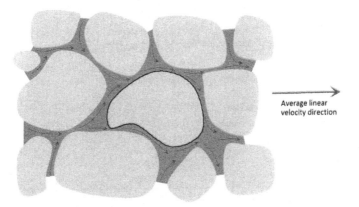

Fig. 1.1 Schematic drawing of a representative elementary volume that encompasses a fluid (shaded) and a solid phase (encircled). These phases may have their distinct temperatures T_s and T_f while the volume as a whole has an average temperature, T

overall extent of the porous medium so that it can be assigned a unique fluid velocity, pressure, and temperature. In quantitative terms, REV is small enough for the first-order Taylor series approximation to be adequate to represent changes taking place across its boundaries. The most significant advantage of using the REV approach is geometrical. The complex pore space is contained within it and need not be specified. What is now relevant is the porosity (ε), namely the ratio of volume of the pore space to that of the porous medium as a whole. At the same time, intensive quantities such as velocity, temperature, pressure drop, and mass fluxes, appropriate for the fluid in the pore space, are replaced by *equivalent* values applicable at the scale of the REV.

Using REV is a volume-averaging approach, except that the volume referred is still small in comparison to the application of interest. It is large enough when compared to the pore scale and contains a combination of fluid and solid phases whose respective proportions are characterized by porosity.

There are consequences to be borne in modeling when the REV approach is adopted. An immediate outcome of volume-averaging is that physical variables such as velocity, pressure, and temperature are defined over fluid volumes that are not clearly delineated. The physical laws that help develop the mathematical model are to be stated in terms of these integrated quantities. The mathematical form of these conservation principles, thus, can only inexactly be written. Equality among the terms of the governing equations can be restored only by the introduction of empirical factors. Such factors can be exactly derived in integral form by going from the fluid flow equations in the pore space to the REV, if the pore geometry is clearly specified. In most applications, the pore space is not easily identified and the analytical derivation of these empirical factors is not possible. An experimental determination is then the only approach available, making laboratory measurements

Fig. 1.2 REV (of the order
of a few hundred microns),
laboratory scale (tens of
millimeters), and the field
scale (1–1000 m) shown in a
comparison plot; d_p is the
particle diameter (10–100 μ)

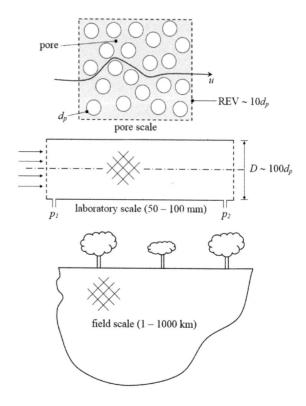

a vital step in the setting up of the governing equations of transport in porous
media.

A comparison of field-, laboratory-, and REV-scales is shown in Fig. 1.2.

1.3   Mathematical Modeling of Fluid Flow

Flow and transport phenomena in porous media are encountered in a great many
fields of engineering: civil, chemical, mechanical, and petroleum, to name a few.
A wide variety of flow patterns can be realized in the porous medium. These are
controlled by the flow regime (for example, steady or unsteady, laminar or turbulent), fluid interfaces in multiphase flow, non-equilibrium phenomena in the
transported variables such as pressure, temperature and concentration, and phase
change. Additionally, it is now well-established that the flow patterns and transport
depend intricately on the structure of the pore space and the solid matrix. In fact, in
a majority of applications, the non-uniformities in the pore geometry play a defining
role in fixing flow and transport.

Strategies employed in the past for studying transport phenomena in a non-uniform porous medium include the following:

1. Formulate the governing equations for an inhomogeneous, anisotropic region. Properties such as permeability and dispersion are interpreted here as second-order tensors that have built-in dependence on spatial location.
2. Permeability and dispersion are treated as random variables with a well-defined mean (that can be a function of space). These are associated with appropriate statistics that account for fluctuations in the physical properties and possible spatial correlation, as well correlation between properties themselves.

While approach 1 leads to a deterministic model that can be solved by a traditional numerical technique, approach 2 leads to a stochastic model that must be solved several times to generate realizations of the resulting flow and scalar fields.

One of the major difficulties encountered in the two approaches referred above pertains to the sensitivity of the numerical solution on the specification of pore-scale variation (deterministically or via statistics) of the macroscopic properties. For example, a refinement of the computational grid will permit greater details of the pore structure to be prescribed and yield a completely new solution. Yet another difficulty is experienced when the model parameters have to be determined from laboratory or field-scale experiments. The scale of measurements may be such that variations on a lower scale are completely masked. The difficulties become compounded when transport occurs over a wide range of length scales or over an extremity of pore scales. The past approach has been to determine parameters in such a way that a few limited goals of the mathematical model are satisfied, though the estimations may not strictly conform to every possible variation in the pore geometry.

A recent approach to characterizing transport in porous media is to view the geometry of the flow path as comprising a hierarchical collection of length scales that can range in size from a few to several orders of magnitude. It includes micro- and nanoscale porous media and the relevant transport mechanisms in them. The mathematical models of flow and transport for such domains have been proposed in the literature, though the subject is in the developmental phase. The idea here is to average the very fundamental governing equations over each ladder of the hierarchy. This has the advantage of retaining the deterministic nature of the governing equations, while considerable changes in the pore structure can be accommodated. For example, flow in a fractally heterogeneous porous medium can be represented by an average permeability and a fractal dimension, rather than a permeability field that is grid-dependent. Thus, the process of averaging the original governing equations over a discrete set of length scales generates a system of equations with a new set of parameters.

1.4   Darcy's Law

While the pore-scale flow of a fluid in a porous region is governed by the Navier–Stokes equations, the governing equation at the scale of a representative elementary volume (REV) is given by Darcy's law

$$\mathbf{u} = -\frac{K}{\mu}\nabla p \tag{1.1}$$

Here, $\mathbf{u}$ is the REV-averaged velocity vector and p is fluid pressure. In addition, K is permeability and μ is the dynamic viscosity of the fluid. To a first approximation, the local pore-scale velocity $\mathbf{u}*$ can be estimated using the porosity of the medium (ε) as

$$\mathbf{u}* = \frac{\mathbf{u}}{\varepsilon}$$

Historically, Darcy's law emerged from laboratory-scale experiments, independent of the Navier–Stokes equations. It was only later, that the equivalence between the two was established when it was shown that the Stokes equations, integrated over a suitable REV, reduces to Darcy's law. Since acceleration terms are neglected here, the validity of Darcy's law can be expected under the creeping flow approximation, namely

$$\mathrm{Re} = \frac{\rho u d_p}{\mu} < 1 \tag{1.2}$$

The Reynolds number here is based on the fluid speed u, fluid properties ρ and μ, and the pore dimension d_p. If the pore geometry shows a degree of regularity, for example, a bed of spheres, the process of integration of the Stokes equation will generate an analytical expression for the permeability. In all other contexts, permeability needs to be estimated from experiments. For the bed of spheres of diameter d and porosity ε, permeability (in units of m^2) can be explicitly written as

$$K = \frac{\varepsilon^3 d^2}{180(1 - \varepsilon)^2} \tag{1.3}$$

In other instances, permeability must be determined from experiments. Here, the factor 180 stems from curve fitting and values between 150 and 180 have been quoted in the literature. To make Eq. 1.1 applicable for parameter estimation from experiments, pressure in the porous medium is interpreted as fluid pressure, measurable by using a wall-mounted pressure transducer. In Eq. 1.1, velocity is REV-averaged; in a uniform porous medium, velocity variations are absent and the REV-averaged velocity can be found as the ratio of volumetric flow rate and cross-sectional area. Symbol **u** appearing in Eq. 1.1 is also called the Darcian velocity.

The pressure–velocity relationship of Darcy's law can be compared with Fourier's law of heat conduction

$$\mathbf{q} = -k\nabla T$$

and the voltage–current relationship of Ohm's law. It can also be seen as a simplified form of the momentum equation of a Newtonian fluid, being analogous to the friction factor relation

$$f = \frac{64}{\mathrm{Re}}$$

for laminar fully developed flow in a straight tubes of circular cross-section.

If higher flow speeds are to be expected, the pore diameter based Reynolds number would exceed unity, and the starting point of analysis is the Navier–Stokes equations. This approach accommodates unsteady effects, formation of eddies, and wall effects. Within the broad framework of Stokes and Navier–Stokes equations, permeability decreases quadratically with the particle diameter. This dependence arises from the no-slip condition between the fluid and the solid phases at the solid boundary.

Historically, Darcy's law has greatly revolutionized analysis in applications such as groundwater flow and has played an influential role in the development of the subject of transport in porous media. Though applicable in a limited context, it has formed the basis of extensions arising from geometrical as well as dynamical factors. It is fitting that these extensions are called non-Darcy models of fluid flow. Governing equation of flow and transport in porous media is discussed in Chap. 2.

The REV-scale formulation of flow and transport creates a new organizing principle wherein a fewer number of variables in a complex geometry is replaced by a larger number of variables and parameters within homogenized space. This viewpoint is illustrated in Fig. 1.3.

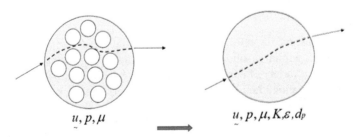

$$u, p, \mu \qquad \qquad u, p, \mu, K, \varepsilon, d_p$$

Fig. 1.3 Schematic drawing of a new organizing principle using REV; the complexity of the pore space on the left is replaced by additional parameters such as permeability K, porosity ε, and particle diameter d_p

1.5   Microscale Phenomena

Liquids, as well as gases, can occupy pores and fractures of physical domains whose length scales extend from a few millimeters down to nanometers. The overall characteristics of the flow and transport processes depend on the connectivity of the pores over the length scales, and the details of the physical processes, as one proceeds from the macro- to the micropores in the porous medium. If there were to be no change in the physical laws underlying transport of mass, momentum, and energy, it would be appropriate to define effective properties for quantities such as permeability, conductivity, and dispersion and treat the porous region as a new continuum composed of REVs. This assumption can, perhaps, be utilized when the pore-scale varies over two or three orders of magnitudes. Whether it continues to hold over 6–7 orders of magnitude is a question that is as yet unresolved. Preliminary research in this direction shows that significant differences are to be anticipated. For example, for very small pore dimensions in the micron regime, effective permeability is greater than what is predicted by Eq. 1.3.

The flow of fluids, fluid–matrix interactions, chemical reactions, wetting characteristics in a nano-environment and slip at a contact line are topics of research. It calls for molecular level experiments as well as simulation to relate microscopic concepts to the macroscopic, such as temperature and pressure. The determination of effective medium properties in the presence of nanoscale phenomena is a source of challenge.

At a small-scale, forces that appear at the gas–liquid interface are considerably large. Pressure drops are substantial and fluid compressibility should be accounted for. In gas flow, the no-slip condition for velocity may be violated while fluid–solid temperature jump may be introduced at the particle surface. Ultimately, the applicability of the continuum equations such as Navier–Stokes is itself in question. The reformulated equations will be have to freshly integrated over the REV to determine the appropriate form of the governing equations for the porous continuum.

1.6   Applications

The terms porous media refer to class of materials that contains connected, fluid-filled, pores within a solid matrix. Such a definition covers a wide range materials used in a variety of engineering applications including, but not limited to, oil and gas exploration, fuel cells and batteries, soil and rock mechanics, and cardiovascular systems. The above applications rely on fluid flow, through the connected pores, often aided by heat and mass transfer phenomena. Understanding flow, heat, mass, and charge transfer through porous media, are, therefore, essential in the analysis of a variety of engineering applications (Fig. 1.4).

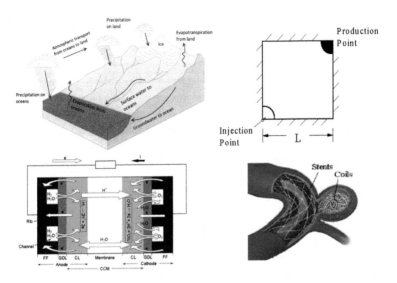

Fig. 1.4 Schematic drawings of applications where porous media modeling is useful for analysis. First row: left—groundwater flow; right—immiscible displacement of oil by water; second row: left—charge transfer in a fuel cell; right—coils embolized in a cerebral artery

While classical idealization based on Darcy's law is primarily limited to empirical formulation, recent research has infused significant rigor in modeling transport phenomena in porous media. Consequently, it is now possible to develop models for several key engineering applications. This monograph links recent advancements to the classical theory leading to the analysis of several emerging applications. The monograph examines four specific applications namely, electrochemical (fuel cell and battery), biomedical (treatment of aneurysm and related ailment), oscillatory flow in regenerators, and geological (methane recovery from marine hydrate) systems. The monograph will act as a valuable resource for the beginner as well as experienced researchers.

1.7 Terminology

The subject of fluid flow and transport phenomena in porous media adopts specific nomenclature that may be explained at the outset. *Porosity* is the ratio of the volume of the pore space, and the total volume that includes the pores and the solid phase. Porosity can be defined at the scale of the entire porous medium or for the REV, for which the term *local porosity* is adopted. The porous region is said to be *saturated* if a single fluid (gas or liquid) fills its pore space. Otherwise, it is *unsaturated*. Here, two or more fluids fill the pore space and may be in motion owing to a superimposed pressure difference. In applications such as groundwater flow, one of the fluids is air and can be taken to be stagnant, at the atmospheric pressure. The second phase, namely water, is mobile and is transported within the porous medium. It is possible for two or more fluids, miscible or immiscible, flowing through the porous medium; this constitutes *multiphase flow*. In a very general setting, the gas phase may be a mixture of air, methane, and water vapor while the liquid phase such as water may have dissolved salts as well as dissolved CO_2. This is an example of multiphase, multicomponent transport in a porous medium. An important approximation often employed in porous media modeling is that of *equilibrium*. If the fluid phases in the pore space are assumed to have equal pressure, the model is said to exhibit mechanical equilibrium. When the fluid phases and the solid medium are all at equal temperatures, the model exhibits thermal equilibrium. The equality of pressure and temperature is local, at the scale of the REV. Globally, pressures and temperatures vary, depending on the flow patterns and the associated boundary conditions. In gas–liquid flows, *interfaces* form in the pore space and will alter the extent of mechanical equilibrium. Thermal non-equilibrium is often seen in rapid transients such as cooling and heating applications. Flow patterns in a porous medium are strongly influenced by its *permeability* (Eq. 1.1) or its distribution in the porous medium. Permeability can be understood as a measure of the ease with flow can taken place in the porous material and plays the role of conductivity in the fluid dynamic context. The limit of an infinite permeability maps the porous region to one containing homogeneous fluid for which the porosity is unity. Hence, the limits

$$\varepsilon \rightarrow 1 \quad K \rightarrow \infty$$

are jointly obtained. Invariably, permeability decreases when interfaces form within the porous medium. The extent of this reduction is specified in terms of *relative permeability* which is unity under saturated flow conditions. Both quantities can vary from point-to-point within the physical domain, REV being the physical point referred here. The discussion of this paragraph presupposes a *continuum approximation* in the fluid medium so that fluid density, viscosity, pressure, and velocity can be uniquely defined. When the pore space becomes excessively small, non-continuum effects can become visible and the fluid is better represented as a collection of particles. These particles may not stick to the sold wall, giving rise to a slip condition in velocity. The mathematical formulation of transport in the non-continuum (mesoscale) domain is briefly described in Chap. 3.

1.8 Closure

The contents that follow are a collection of six major chapters coupled with an introduction (Chap. 1) and a closure (Chap. 8). Two of the six chapters will explain the underlying theories while the remaining four will focus on new applications. Since porous media transport is essentially a multiscale process, the theory of transport phenomena in porous media (the first two chapters) will cover both continuum- and mesoscale phenomena. While continuum formulation will impart further rigor to the empirical porous media models, the mesoscopic model will focus on the physical processes within the pores. The strengths and weaknesses of existing models, as well as future research directions, will be discussed.

Major chapters of the monograph cover four different applications. These are related to electrochemical (fuel cells and batteries), biological (treatment of aneurysm and related ailments), fluid–solid energy exchange in a regenerator, and geological (methane recovery from marine hydrate) systems. The discussion will emphasize the connection between theory and applications. For each of the applications, recent advancements in the literature are described and future research directions identified.

Bibliography

Modeling Flow and Transport in Porous Media

1. J. Bear, Y. Bachmat, *Introduction to Modeling of Transport Phenomena in Porous Media* (Kluwer, Dordrecht, 1990)
2. S. Borzooei, *Contaminant Transport Modeling in Heterogeneous Porous Media: Comparison of Analytical, Numerical and Stochastic Methods in One-dimensional Contaminant Transport Modeling* (Lambert Academic Publishing, Saarbrücken, 2014)
3. F. Civan, *Porous Media Transport Phenomena* (Wiley, Weinheim, 2011)
4. C.K. Ho, S.W. Webb (eds.), *Gas Transport in Porous Media*, Theory and Applications of Transport in Porous Media (Springer, New York, 2010)
5. F.A. Coutelieris, J.M.P.Q. Delgado, *Transport Processes in Porous Media* (Springer, New York, 2012)
6. G. Dagan, Theory of solute transport by groundwater. Annu. Rev. Fluid Mech. **19**, 183–215 (1987)
7. R. de Boer, *Theory of Porous Media: Highlights in Historical Development and Current State* (Springer, New York, 2011)
8. M.J.S. de Lemos, *Turbulence in Porous Media: Modeling and Applications*, 2nd edn. (Elsevier, Amsterdam, 2012)
9. H.-J.G. Diersch, *Finite Element Modeling of Flow, Mass, and Heat Transport in Porous and Fractured Media* (Springer, New York, 2014)
10. P. Dietrich, R. Helmig, M. Sauter, H. Hötzl (eds.), *Flow and Transport in Fractured Porous Media* (Springer, New York, 2010)
11. A. Hunt, R. Ewing, B. Ghanbarian, *Percolation Theory for Flow in Porous Media* (Springer, New York, 2014)
12. Y. Ichikawa, A.P.S. Selvadurai, *Transport Phenomena in Porous Media: Aspects of Micro/Macro Behavior* (Springer, New York, 2014)
13. J.D. Jansen, *A Systems Description of Flow Through Porous Media* (Springer, New York, 2013)
14. M. Kaviany, *Principles of Heat Transfer in Porous Media (Mechanical Engineering Series)* (Springer, New York, 1991)

15. A.H. Kim, R.A. Guyer (eds.), *Nonlinear Elasticity and Hysteresis: Fluid-Solid Coupling in Porous Media* (Wiley-VCH, Weinheim, 2015)
16. O. Kolditz, H. Shao, W. Wang, S. Bauer (eds.), *Thermo-Hydro-Mechanical-Chemical Processes in Fractured Porous Media: Modeling and Benchmarking: Closed-Form Solutions* (Springer, New York, 2014)
17. D. Kulasiri, *Stochastic Dynamics: Modeling Solute Transport in Porous Media* (North Holland, Amsterdam, 2002)
18. D. Kulasiri, *Non-fickian Solute Transport in Porous Media: A Mechanistic and Stochastic Theory*, Advances in Geophysical and Environmental Mechanics and Mathematics (Springer, New York, 2013)
19. D. Long, M.A. Lovell, J.G. Rees, C.A. Rochelle (Eds.), Sediment-hosted gas hydrates: new insights on natural and synthetic systems. Geol. Soc. London (2009)
20. A. Narasimhan, *Essentials of Heat and Fluid Flow in Porous Media* (CRC Press, Boca Raton, 2012)
21. I. Pop, *Convective Heat Transfer: Mathematical and Computational Modeling of Viscous Fluids and Porous Media* (Pergamon, Oxford, 2001)
22. M. Sahimi, *Flow and Transport in Porous Media and Fractured Rock: From Classical Methods to Modern Approaches* (Wiley-VCH, Weinheim, 2011)
23. W.T. Sha, *Novel Porous Media Formulation for Multiphase Flow Conservation Equations* (Cambridge University Press, Cambridge, 2014)
24. Y. Su, J.H Davidson, *Modeling Approaches to Natural Convection in Porous Media* (Springer, New York, 2015)
25. A. Szymkiewicz, *Modeling Water Flow in Unsaturated Porous Media: Accounting for Nonlinear Permeability and Material Heterogeneity* (Springer, New York, 2014)
26. M.A.D. Viera, P. Sahay, M. Coronado, A.O. Tapia (eds.), *Mathematical and Numerical Modeling in Porous Media: Applications in Geosciences*, Multiphysics Modeling (CRC Press, Boca Raton, 2012)
27. S. Whitaker, *The Method of Volume Averaging (Theory and Applications of Transport in Porous Media)* (Springer, New York, 2010)
28. D. Zhang, *Stochastic Methods for Flow in Porous Media: Coping with Uncertainties* (Academic Press, Cambridge, 2001)

Multiphase Flow in Porous Media

1. D.B. Das, S.M. Hassanizadeh (eds.), *Upscaling Multiphase Flow in Porous Media: From Pore to Core and Beyond* (Springer, New York, 2010)
2. R.E. Ewing, *The Mathematics of Reservoir Simulation* (SIAM, Philadelphia, 1983)
3. W.E. Fitzgibbon (ed.), *Mathematical and Computational Methods in Siesmic Exploration and Reservoir Modelling* (SIAM, Philadelphia, 1986)
4. A. Oechsner, G.E. Murch (eds.), *Heat Transfer in Multi-Phase Materials* (Springer, New York, 2011)
5. G.F. Pinder, W.G. Gray, *Essentials of Multiphase Flow in Porous Media* (Wiley, Weinheim, 2008)

Biomedical Modeling in Porous Media

1. M. Izadifar, *A Porous Media Approach for Physiological Modeling: Mathematical Modeling, Dynamic Simulation and Sensitivity Analysis of Drug Elimination by the Liver* (LAMBERT Academic Publishing, Saarbrücken, 2013)

2. P. Vadasz, *Emerging Topics in Heat and Mass Transfer in Porous Media: From Bioengineering and Microelectronics to Nanotechnology*, Theory and Applications of Transport in Porous Media (Springer, New York, 2010)
3. K. Vafai, *Porous Media: Applications in Biological Systems and Biotechnology* (CRC Press, Boca Raton, 2011)

Numerical Techniques in Porous Media

1. P. Bastian, Numerical computation of multiphase flows in porous media, Heidelberg, June 1999
2. Z. Chen, R.E. Ewing, Z.-C. Shi, *Numerical Treatment of Multiphase Flows in Porous Media: Proceedings of the International Workshop Held at Beijing, China, 2–6 August 1999*, Lecture Notes in Physics (Springer, New York, 2010). Reprint of original 1st ed. 2000
3. Z. Chen, G. Huan, Y. Ma, *Computational Methods for Multiphase Flows in Porous Media*, Computational Science and Engineering (SIAM, Philadelphia, 2006)
4. J.M. Crolet (ed.), *Computational Methods for Flow and Transport in Porous Media*, Theory and Applications of Transport in Porous Media (Springer, New York, 2010). reprint of 1st ed 2000
5. J.M.P.Q. Delgado, B. de Lima, A. Gilson, M.V. da Silva (eds.), *Numerical Analysis of Heat and Mass Transfer in Porous Media* (Springer, New York, 2012)
6. T. Ertekin, J.H. Abou-Kassem, G.R. King, *Basic Applied Reservoir Simulation*, vol. 10 (SPE Textbook, 2001
7. J.A. Ferreira, S. Barbeiro, G. Pena, M.F. Wheeler (eds.), *Modeling and Simulation in Fluid Dynamics in Porous Media*, Springer Proceedings in Mathematics & Statistics (Springer, New York, 2015)
8. F. Kamyabi, *Multiphase Flow in Porous Media: Numerical Approach for Modeling Multiphase Flow* (Lambert Academic Publishing, Saarbrücken, 2014)
9. A. Öchsner, G.E. Murch, M.J.S. de Lemos (eds.), *Cellular and Porous Materials: Thermal Properties Simulation and Prediction* (Wiley-VCH, Weinheim, 2008)
10. S. Tyson, *An Introduction to Reservoir Modeling* (Pipers' Ash, 2007)

Experiments in Porous Media

1. R. de Boer (ed.), *Porous Media: Theory and Experiments* (Springer, New York, 2014). 1999 first edition
2. W. Ehlers, J. Bluhm (eds.), *Porous Media: Theory, Experiments and Numerical Applications* (Springer, New York, 2010)

Hierarchical Modeling

1. J.H. Cushman, *The Physics of Fluids in Hierarchical Porous Media: Angstroms to Miles* (Kluwer Academic, Dordrecht-Boston, 1997)

Turbulent Flow

1. Y. Takatsu, T. Masuoka, Turbulent phenomena in flow through porous media. J. Porous Media **1**, 243–251 (1998)
2. D. Getachew, W.J. Minkowycz, J.L. Lage, A modified form of the k-ε model for turbulent flow of an incompressible fluid in porous media. Int. J. Heat Mass Transfer 2909–2915 (2000)
3. M.J.S. de Lemos, M.H.J. Pedras, Recent mathematical models for turbulent flow in saturated rigid porous media. ASME Fluids Eng. **123**(4), 935–940 (2001)
4. K. Suga, Understanding and modelling turbulence over and inside porous media. Flow Turbul. Combust. **96**(3), 717–756 (2016)

Chapter 2
Equations Governing Flow and Transport in Porous Media

K. Muralidhar

In the simplest situation of a large bed of small rigid spheres of uniform diameter, flow in the pore space can be represented by Darcy's law [1]. As discussed in Chap. 1, this relationship, stated at the scale of the representative elementary volume (REV), arises from volume-averaging Navier–Stokes equations at the low Reynolds number limit in a repeating geometry [2, 3]. Historically, Darcy's law was proposed as an empirical relationship on the basis of laboratory experiments while the theoretical basis was developed in later years. In this respect, the mass, momentum, and energy equations in a porous medium are posed as mathematical models rather than natural laws. This spirit of empirical–theoretical formulation of flow and transport in porous media has pervaded analysis and will be followed in the rest of the chapter.

With burgeoning applications in porous media, a wide variety of possibilities may be realized during flow and transport. While single and multiphase flows in porous media are discussed in the following sections, several other specialized topics such as radiative transport and solid–fluid phase change have not been covered [4–6]. The discussion here is restricted to devices and processes in which one or more fluids pass through the porous region, exchanging energy and mass with the solid phase.

2.1 Darcy's Law

Originally stated for a bed of spheres, Darcy's law for fluid flow in the pore space is given by the relationship

$$\mathbf{u} = -\frac{K}{\mu}\nabla p \qquad (2.1)$$

Here, $\mathbf{u}$ is the fluid velocity obtained as an average over the REV, and p is fluid pressure at the pore scale. The symbol K represents the medium permeability, and

© Springer International Publishing AG 2018
M.K. Das et al., *Modeling Transport Phenomena in Porous Media with Applications*, Mechanical Engineering Series,
https://doi.org/10.1007/978-3-319-69866-3_2

μ is the dynamic viscosity of the fluid. The porous medium is taken to be *saturated* with the fluid of interest in the sense that fluid–fluid interfaces do not form, and a single fluid prevails in the pore space. Let d_p be the particle size and U, a velocity scale. Equation 2.1 has been found to be applicable for Reynolds number of up to unity, namely [1–3, 6, 7]

$$\text{Re} = \frac{\rho U d_p}{\mu} < 1$$

The permeability of a porous medium is a property that depends on the pore size and the pore structure. The fact that permeability may vary from one geometry to another makes the applicability of Eq. 2.1 quite restrictive. On the other hand, dimensional analysis shows permeability to be a function of porosity ε and particle diameter d_p, each representing the pore geometry and pore size, respectively. The Carman–Kozeny relationship connects these quantities empirically but with dimensional correctness as [2, 8, 9]

$$K = \frac{d_p^2 \varepsilon^3}{180(1 - \varepsilon)^2} \tag{2.2}$$

With the particle diameter expressed in meters, permeability has dimensions of m^2. In applications, one can expect to have a length scale for the device itself, say L, and the ratio defined as

$$Da = \frac{K}{L^2} \tag{2.3}$$

is called the Darcy number. In many applications, the particle diameter is of the order of a fraction of a millimeter while the device length scale is around a meter or beyond. Naturally, Darcy numbers of the order of 10^{-6} and smaller and quite common. Larger Darcy numbers will indicate larger particle diameter and pore size. The resulting Reynolds number will be greater than unity leading to a violation of Darcy's law. Hence, Eq. 2.1 can be properly stated as being applicable in the limit of $Da < < 1, \text{Re} < 1$.

The Carman–Kozeny relationship for permeability has limited applicability. Even for a regular bed of spheres, the numerical factor shows a spread over 150–180. It has the advantage of representing the flow space in terms of porosity while the flow size is referred to the particle diameter. These small-scale features are then projected on to the size of the REV. In more complex systems such as meshes and an array of cylinders that are assumed to behave like porous media, an explicit permeability–porosity relationship may not exist. Equation 2.1 must then be used to determine permeability from a calibration experiment where pressure drop and velocity are individually measured [10].

Darcy's law (Eq. 2.1) is remarkably similar to Fourier's law of heat conduction where the heat flux and temperature gradient are connected via thermal conductivity. Heat transfer across a finite temperature difference is thermodynamically irreversible with a local dissipation rate proportional to the product of temperature gradient and heat flux. Analogously, flow through porous media is irreversible, and the power required to sustain flow is proportional to the product of velocity and pressure gradient. The power thus supplied is internally dissipated as thermal energy through the action of viscosity.

Darcy's law is most appropriately used to determine fluid velocity and the flow rate, when the pressure gradient is known. For example, flow in a tube filled with a porous medium will experience a pressure drop of $p_2 - p_1 = \Delta p$ over a length L between points 1 and 2. The average fluid velocity in the tube along its axis can now be calculated as

$$u = -\frac{K}{\mu}\left(\frac{\Delta p}{L}\right) = \frac{K}{\mu}\frac{p_1 - p_2}{L}$$

Volume flow rate is obtained as a product of velocity and the cross-sectional area. The power consumed to sustain flow is the product of pressure drop and volume flow rate.

In a multidimensional flow field, the local pressure gradient is also an unknown. Here, Darcy's law is viewed as a momentum equation wherein pressure gradient balances viscous forces at the fluid–particle boundary. Other contributions such as body forces, inertia, and unsteadiness are set to zero. The momentum equation is then to be supplemented by the mass balance equation. Assuming flow to be incompressible, mass balance equation at the scale of the REV is given as [4–9, 11]:

$$\nabla \cdot \mathbf{u} = 0 \tag{2.4}$$

Equations (2.1) and (2.4) can be jointly used to obtain the flow and pressure distribution in the porous medium.

To a first approximation, Darcy's law is applicable without change during flow transients as well. The justification here is that the characteristic length scale related to the duration of the transient, namely d_p, the particle diameter, is small. Consequently, the timescale

$$\tau = \frac{d_p}{U}$$

is also a small quantity. In applications, the timescale over which changes in pressure and velocity take place are usually much larger. For conditions where flow or pressure is driven in time, Darcy's law (Eq. 2.1) is now taken to be applicable for each time instant.

Between velocity and pressure, velocity can be eliminated using the mass balance equation, yielding a single equation for pressure, namely

$$\nabla^2 p = 0 \tag{2.5}$$

The pressure equation can be solved along with suitable boundary conditions. Combined with Eq. 2.1, the velocity field can be mapped in the porous region.

2.1.1 Cartesian and Cylindrical Coordinate Systems

Equation 2.5 is in coordinate-free form; specific forms of the pressure equation in Cartesian and cylindrical coordinate systems are given as follows:

$$\text{Cartesian} \quad \frac{\partial^2 p}{\partial x^2} + \frac{\partial^2 p}{\partial y^2} + \frac{\partial^2 p}{\partial z^2} = 0$$

$$\text{Cylindrical} \quad \frac{\partial^2 p}{\partial r^2} + \frac{1}{r}\frac{\partial p}{\partial r} + \frac{1}{r^2}\frac{\partial^2 p}{\partial \theta^2} + \frac{\partial^2 p}{\partial z^2} = 0 \tag{2.6}$$

The velocity components are determined using Eq. 2.1 as:

$$\text{Cartesian} \quad u = -\frac{K}{\mu}\frac{\partial p}{\partial x} \quad v = -\frac{K}{\mu}\frac{\partial p}{\partial y} \quad w = -\frac{K}{\mu}\frac{\partial p}{\partial z}$$

$$\text{Cylindrical} \quad u^r = -\frac{K}{\mu}\frac{\partial p}{\partial r} \quad u^\theta = -\frac{1}{r}\frac{K}{\mu}\frac{\partial p}{\partial \theta} \quad u^z = -\frac{K}{\mu}\frac{\partial p}{\partial z} \tag{2.7}$$

Equations 2.5–2.6 can be solved by the tools of classical analysis [12–14], for example, separation of variables in finite domains of regular shape and transform techniques in semi-infinite geometries. They require prescription of boundary conditions over the entire boundary in one of the following standard forms:

$$\text{Dirichlet: } p(x, y, z) \text{ specified}$$

$$\text{Neumann: } \frac{\partial p}{\partial n} \text{ specified}$$

$$\text{Robin: } ap(x, y, z) + b\frac{\partial p}{\partial n} \text{ specified (constants } a \text{ and } b) \tag{2.8}$$

For example, at an injection or fluid production point, pressure is specified. On a no-flow, impermeable boundary, the normal derivative of pressure is zero. The third category of boundary condition is also possible. The specified quantity can be zero, a constant, or a function of time.

Some porous media problems can be formulated with volumetric sources (of mass) distributed as $Q(\mathbf{x})$ (units of m^3/s per m^3 volume of REV) for which the mass balance equation is written as

$$\nabla \cdot \mathbf{u} = Q$$

Combined with Darcy's law, the pressure equation takes on the form

$$\nabla^2 p + \frac{\mu}{K} Q = 0$$

The boundary conditions continue to be given by Eq. 2.8.

Since the pressure gradient along a solid wall is nonzero, the tangential component of wall velocity is nonzero and the no-slip condition does not hold. For an impermeable wall, the normal derivative of pressure is zero, and hence, the normal velocity is also zero. The inapplicability of the no-slip velocity condition is peculiar to the Darcy formulation of flow in porous media. The non-Darcy system of equations (Sect. 2.4) permits the no-slip condition and is more realistic.

For a clear fluid medium lying adjacent to a porous medium, the boundary conditions for Navier–Stokes equations are modified to account for slip at the interface. In place of the no-slip condition, the following Beavers–Sparrow condition [9] is utilized:

$$\frac{\partial u_f}{\partial y} = \frac{\alpha_{BJ}}{\sqrt{K}} (u_f - u_{PM})$$

Here, suffix 'f' refers to the clear fluid medium, PM is the porous medium, and α_{BJ} is an empirical (Beavers–Joseph) parameter. The Dirichlet boundary condition for velocity is now replaced by a Neumann condition. It can be expected that the normal velocity component will remain continuous across the clear–porous interface. The governing equations in the porous medium and the clear fluid are thus coupled at their interface.

2.1.2 Inhomogeneous Media

Here, the pore space dimensions may vary from one location to another. If an average pore dimension can be identified at the scale of the REV, one can define a permeability function $K(\mathbf{x})$ and the corresponding form of Darcy's law

$$\mathbf{u} = -\frac{K(\mathbf{x})}{\mu} \nabla p \tag{2.9}$$

Combining with the mass balance equation, Eq. 2.9 can be expressed in terms of pressure as

$$\nabla \cdot \{K(\mathbf{x})\nabla p\} = 0 \tag{2.10}$$

The pressure equation is solved along with the Dirichlet, Neumann, or Robin boundary conditions described by Eq. 2.8. Equation 2.10 expressed in Cartesian coordinates is of the form

$$\frac{\partial}{\partial x}\left[K(x,y,z)\frac{\partial p}{\partial x}\right] + \frac{\partial}{\partial y}\left[K(x,y,z)\frac{\partial p}{\partial y}\right] + \frac{\partial}{\partial z}\left[K(x,y,z)\frac{\partial p}{\partial z}\right] = 0$$

2.1.3 Anisotropic Media

Darcy's law (Eq. 2.1) can be expressed in index notation as

$$u_i = -K\frac{\partial p}{\partial x_i} \tag{2.11}$$

Here, permeability is a scalar function that is a constant or a prescribed function of position, as in Eq. 2.9. Equation 2.11 also admits tensor forms of permeability wherein permeability is a Cartesian second-order tensor, namely

$$u_i = -K_{ij}\frac{\partial p}{\partial x_j} \tag{2.12}$$

Equation 2.12 expands in a Cartesian coordinate system as

$$\begin{aligned}
u &= -K_{11}\frac{\partial p}{\partial x} - K_{12}\frac{\partial p}{\partial y} - K_{13}\frac{\partial p}{\partial z} \\
v &= -K_{21}\frac{\partial p}{\partial x} - K_{22}\frac{\partial p}{\partial y} - K_{23}\frac{\partial p}{\partial z} \\
w &= -K_{31}\frac{\partial p}{\partial x} - K_{32}\frac{\partial p}{\partial y} - K_{33}\frac{\partial p}{\partial z}
\end{aligned} \tag{2.13}$$

Here, K_{ij} is a permeability tensor and has nine components, with indices 'i' and 'j' running over 1...3. These respectively refer to the Cartesian coordinate directions x, y, and z. Combining with the continuity equation, Eq. 2.13 can be cast in a form suitable for pressure calculation as [15]

$$\begin{aligned}
&\frac{\partial}{\partial x}\left[K_{11}\frac{\partial p}{\partial x} + K_{12}\frac{\partial p}{\partial y} + K_{13}\frac{\partial p}{\partial z}\right] + \frac{\partial}{\partial y}\left[K_{21}\frac{\partial p}{\partial x} + K_{22}\frac{\partial p}{\partial y} + K_{23}\frac{\partial p}{\partial z}\right] \\
&+ \frac{\partial}{\partial z}\left[K_{31}\frac{\partial p}{\partial x} + K_{32}\frac{\partial p}{\partial y} + K_{33}\frac{\partial p}{\partial z}\right] = 0
\end{aligned} \tag{2.14}$$

In homogeneous media, the permeability tensor is symmetric, and the nine components reduce to six. In isotropic and homogeneous media, the nine components of the permeability tensor reduce to one, making permeability a scalar quantity. It is worth repeating here that Darcy's law as given by Eq. 2.1 is applicable for a homogeneous and isotropic medium, characterized by a scalar permeability.

In an orthotropic medium, the off-diagonal components of permeability are zero while the principal components of permeability are nonzero and distinct. Here

$$K_{ij} = \begin{bmatrix} K_{11} & 0 & 0 \\ 0 & K_{22} & 0 \\ 0 & 0 & K_{33} \end{bmatrix}$$

The pressure equation can now be derived as

$$\frac{\partial}{\partial x}\left(K_{11}\frac{\partial p}{\partial x}\right) + \frac{\partial}{\partial y}\left(K_{22}\frac{\partial p}{\partial y}\right) + \frac{\partial}{\partial z}\left(K_{33}\frac{\partial p}{\partial z}\right) = 0 \qquad (2.15)$$

The permeability components K_{ii} can be functions of position or take on purely constant values. For a homogeneous and isotropic medium,

$$K_{11} = K_{22} = K_{33} = K$$

For anisotropic media, one can identify principal directions with respect to which the off-diagonal components of permeability are zero. The Cartesian coordinate system can then be aligned with the principal directions of the porous medium being studied, leading to some simplification in the governing equations.

2.1.4 Compressible Flow

For a start, think of a porous medium whose pore space does not alter with pressure, so that it is characterized by a constant porosity. In certain circumstances, the fluid filling the pore space may have to be treated as compressible. The mass balance equation for a compressible fluid in a fixed pore space is now expressed as

$$\varepsilon\frac{\partial \rho}{\partial t} + \nabla \cdot \rho\mathbf{u} = 0 \qquad (2.16)$$

Here, ρ is fluid density, ε is porosity, and $\mathbf{u}$ is the REV-averaged fluid velocity. With mass sources, the right-hand side is replaced by $\rho Q(\mathbf{x})$. Eq. 2.16 follows the mass balance equation of a compressible fluid of homogeneous fluid-filled porous media, except that the first term includes porosity. This correction arises from the fact that the amount of fluid contained in the REV is a fraction of the volume of REV, as defined by the porosity.

If the flow rates are small enough for the pore-scale Reynolds number to be less than unity, Stokes equations will hold in the fluid phase, and its integrated form at the scale of the REV, namely Darcy's law (Eq. 2.1), can once again be used as a momentum equation. Hence, working with a homogeneous and isotropic medium, velocity and pressure are connected as

$$\mathbf{u} = -\frac{K}{\mu}\nabla p$$

The system of equations can be closed when an equation of state relating density and pressure is prescribed. For gases, changes in density will be accompanied by those in pressure and temperature. Accordingly, an energy equation is added to the system of equations governing compressible flow of gas in a porous medium. For a liquid, density may be connected to pressure through compressibility S as

$$S = \frac{1}{\rho}\frac{\partial \rho}{\partial p}$$

Here, S is a positive quantity (units of $(N/m^2)^{-1}$) and is a constant over a range of fluid pressures. Combining the unsteady mass balance equation with Darcy's law, we get

$$S\rho\varepsilon\frac{\partial p}{\partial t} = \nabla \cdot \left(\rho\frac{K}{\mu}\nabla p\right)$$

A simplified form of the pressure equation that is adequate for mildly compressible fluids ($\rho \sim$ constant) can be derived by canceling density on both sides. It reads

$$\frac{\partial p}{\partial t} = \frac{K}{\mu S\varepsilon}\nabla^2 p \tag{2.17}$$

If the porous medium is also compressible, S can be interpreted as the combined compressibility of the medium and the fluid within. Equation 2.17 shows that the role of compressibility is to sustain flow transients in a porous medium.

The unsteady pressure equation for a compressible liquid in an incompressible porous medium has a form analogous to the diffusion equation of heat transfer. Analytical solutions developed in specialized texts related to conduction heat transfer can be extended to the present context [14].

Compressibility of the fluid is important for gaseous media. It is also important when liquids are pumped at high pressures through a porous reservoir. For small pore dimensions, permeability (and hence, Darcy number) is small, pressure drop is large, and compressibility will once again be relevant.

If the porous medium carried gas, the compressibility relationship emerges (to a first approximation) via the ideal gas law

$$\frac{p}{\rho} = RT$$

For near-isothermal conditions, we may write $\rho = Cp$ for the ideal gas law leading to the nonlinear governing equation

$$\frac{\partial p}{\partial t} = \nabla \cdot \left(\frac{K}{\mu \varepsilon} \nabla p^2 \right) \qquad (2.18)$$

For steady state, Eq. 2.18 is linear in the new variable—square of pressure. For constant properties, Eq. 2.18 simplifies to

$$\nabla \cdot \nabla p^2 = 0$$

2.1.5 Effect of Gravity

In large-scale reservoirs, gravity forces become relevant and alter the flow field. Gravity-related corrections will appear in the momentum equation, namely Darcy's law in the form

$$u = -\frac{K}{\mu}(\nabla p + \rho g) = -\frac{K}{\mu}(\nabla \tilde{p}) \qquad (2.19)$$

Here $\tilde{p} = p + \rho g z$. Symbol z stands for the vertical direction opposed to the direction of gravity. The augmented pressure variable can be interpreted as a gravity-modified pressure. Equations derived earlier carry over unchanged with p replaced by $\tilde{p}$, when gravitational effects are to be accounted for.

2.2 Brinkman-Corrected Darcy's Law

Consider the simplest layout of flow in a parallel plate channel of length L and opening $2H$, filled with homogeneous isotropic porous media of permeability K, driven by an overall pressure difference. The governing equation for pressure reduces to

$$\frac{d^2 p}{dx^2} = 0$$

The pressure variation is thus linear in the flow direction; using Darcy's law, the velocity components are evaluated as

$$u = -\frac{K}{\mu}\frac{p_2 - p_1}{L}; \quad v = 0$$

The velocity components predicted by Darcy's law are seen here to be independent of the axial coordinate and hence, *fully developed*. When the pressure difference changes with time, the *x*-component of velocity will change instantaneously, without any time lag. These results are not unexpected because the characteristic length scale that determines transient duration as well as the flow development is the pore diameter, which is a small quantity, relative to H and L.

A second consequence of a constant velocity component is that it is spatially uniform in the cross-sectional *y*-direction, including the wall. This solution violates the no-slip condition between a fluid and a solid boundary commonly seen in continuum flow applications. The interpretation here is that porous media lead to extremely thin wall boundary layers. The spatially uniform velocity solution is then applicable in an experiment over most of the channel width. Based on dimensional reasoning, one can expect the boundary-layer thickness to scale with $\sqrt{K}$, and so with the square root of Darcy number $\sqrt{Da} = \sqrt{K}/H$, where H is the channel half-width. Thus, the constant velocity solution is expected to be realized in dense, low permeability porous media. Here, pressure drop arises mainly from skin friction within the body of the porous medium, unlike homogeneous fluids where the origin of pressure drop is skin friction at the solid walls.

The uniform velocity solution is expected to fail in moderate and high permeability porous regions confined in a channel or a tube. An additional factor that needs consideration is the variation of permeability of a porous region near a solid wall. As shown in Fig. 2.1, porosity near a wall is greater in comparison to the bulk, leading to higher permeability. The wall boundary layers are thickened, and the entire velocity profile can now be obtained only after accounting for the no-slip condition.

A useful framework for introducing the no-slip condition in Darcy's law is available through the Brinkman's approximation which introduces a viscous shear stress term in Darcy's law. The Brinkman-corrected Darcy's law is written as

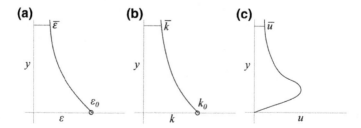

Fig. 2.1 a–b Wall porosity and permeability variation in a channel; **c** schematic drawing of velocity variation in the near-wall region; peak porosity and permeability are indicated as ε_0 and K_0 respectively

$$-\nabla p - \frac{\mu}{K} u + \frac{\tilde{\mu}}{\varepsilon} \nabla^2 \mathbf{u} = 0 \tag{2.20}$$

Comparing with Eq. 2.1, the correction is seen in the third term involving an effective viscosity $\tilde{\mu}$. Given the overall uncertainty in model parameters such as K and ε, the effective viscosity is approximated as the fluid viscosity itself [8, 9].

Since Eq. 2.20 is a second-order partial differential equation in velocity, the no-slip condition can now be conveniently introduced at the walls. A sample solution of velocity distribution is sketched in Fig. 2.1. The overall pressure drop will arise from shear stresses at the liquid–particle interface of the porous region as well as stresses at the liquid–channel boundary that enforce the no-slip condition. It may be noted in Fig. 2.1 that the near-wall velocity variation has a local maximum arising from larger porosity and permeability, a phenomenon called *channeling*. It refers to preferential flow in the near-wall region owing a larger pore scale in comparison to the bulk of the porous medium.

2.3 Forschheimer-Extended Darcy's Law

Darcy's law requires correction when the particle-diameter-based Reynolds number exceeds unity. Such a flow field is commonly encountered in high porosity systems where the fluid velocities are high. In gas flows, viscosity is low, and Reynolds numbers can once again exceed unity.

The immediate consequence of increasing Reynolds number is that form drag experienced by the particles of the solid phase becomes larger in relation to viscous drag. In the limiting case of high Reynolds numbers, form drag can be substantially larger than viscous drag. Figure 2.2 shows the variation of drag coefficient C_D of a sphere as a function of Reynolds number Re_d. With D as total drag,

$$C_D = \frac{D}{\frac{1}{2}\rho u^2 A_p} \quad \text{and} \quad Re_d = \frac{\rho u d}{\mu}$$

Here, ρ is fluid density and A_p is a reference area, usually chosen as the projected area of the object normal to the flow direction. For a sphere of diameter d,

$$A_p = \pi d^2$$

For small Reynolds numbers, drag coefficient decreases with Re_d, giving rise to the following scaling relation:

$$C_D = \frac{D}{\frac{1}{2}\rho u^2 A_p} = \frac{A}{Re_d}. \text{ Hence, drag varies as } D = \text{constant} \times u$$

For larger Reynolds numbers ($20 < Re_d < 10{,}000$), Fig. 2.2 shows near constancy of the drag coefficient. Hence

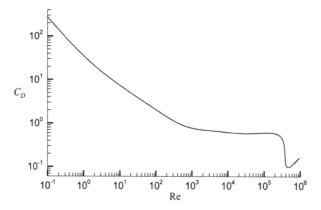

Fig. 2.2 Variation of drag coefficient with Reynolds number for a sphere placed in uniform approach flow

$$C_D = \frac{D}{\frac{1}{2}\rho u^2 A_p} = \text{constant} \sim 1; \quad D = \text{constant} \times u^2$$

Combining the two limits, one can write the following expression for drag experienced by a particle of the porous medium:

$$D = au + bu^2$$

The square-law dependence arises from an adverse pressure gradient and flow separation at the surface the sphere. The quadratic trend realized for a single sphere can be extended to a porous medium comprising a bed of spheres. Experiments indeed show that a porous medium made up of small particles will show drag scaling linearly with velocity at a low Reynolds number and square of velocity for $\text{Re} \gg 1$. The low Reynolds number limit of drag proportional to velocity translates as Darcy's law (Eq. 2.1). An extended form of Darcy's law valid for a wider range Reynolds numbers can be written analogously to drag coefficient as

$$\frac{\Delta p}{L} = -\frac{\mu}{K}u - \frac{\rho F}{\sqrt{K}}u^2$$

The symbol Δp represents pressure drop and is negative with respect to the flow direction. Here, F is a parameter to be determined from experiments. For a bed of uniform-diameter rigid spheres, the following correlation has been reported in the literature [5, 8, 9]:

$$F = \frac{1.8}{(180\varepsilon^5)^{0.5}}\varepsilon$$

In applications, it will have to be experimentally determined from calibration experiments; see Chap. 6 for an example. The square-law correction of Darcy's law is called the Forschheimer term, F being the Forschheimer constant. Darcy's law generalized to accommodate a wider range of Reynolds numbers is expressed as

$$-\nabla p - \frac{\mu}{K}\mathbf{u} - \frac{\rho F}{\sqrt{K}}\mathbf{u}|\mathbf{u}| = 0 \qquad (2.21)$$

The modulus of velocity ensures that the formulation is correct for positive as well as negative flow directions.

The Brinkman and Forschheimer-corrected Darcy's law is written as

$$-\nabla p - \frac{\mu}{K}\mathbf{u} - \frac{\rho F}{\sqrt{K}}\mathbf{u}|\mathbf{u}| + \frac{\mu}{\varepsilon}\nabla^2\mathbf{u} = 0 \qquad (2.22)$$

Equation 2.22 is a form of a momentum equation in which a balance of fluid pressure, viscous forces at the fluid–particle boundary, viscous forces within the fluid medium, generally arising at the confining walls, and form drag is attained at the scale of the REV.

2.4 Non-darcy Model of Flow

Equation 2.22 extends Darcy's law by including a variety of external forces. Indeed, this approach can be used to bring in additional forces arising from, for example, buoyancy and electromagnetism. Equation 2.22 requires the sum of all external forces to be zero, a condition applicable only for non-accelerating flow fields. The expression for fluid acceleration a in a Lagrangian frame of reference can be written as

$$\mathbf{a} = \frac{1}{\varepsilon}\frac{d\mathbf{u}}{dt}$$

Here, the symbol $\mathbf{u}$ continues to represent the REV-averaged velocity, with $\mathbf{u}/\varepsilon$ being the local fluid velocity vector. Hence, acceleration referred here is that of the fluid alone, while the solid phase is stationary.

It is useful to employ the Eulerian representation of acceleration [11]

$$\mathbf{a} = \frac{1}{\varepsilon}\frac{d\mathbf{u}}{dt} = \frac{1}{\varepsilon}\left(\frac{\partial\mathbf{u}}{\partial t} + \frac{1}{\varepsilon}\mathbf{u}\cdot\nabla\mathbf{u}\right)$$

Temporal and spatial acceleration terms are now explicitly written out. Equating the sum of all external forces to mass × acceleration, Eq. 2.22 can be generalized to the form

$$\frac{\rho}{\varepsilon}\frac{d\mathbf{u}}{dt} = -\nabla p - \frac{\mu}{K}\mathbf{u} - \frac{\rho F}{\sqrt{K}}\mathbf{u}|\mathbf{u}| + \frac{\mu}{\varepsilon}\nabla^2\mathbf{u}$$

Here, ρ is fluid density. Hence

$$\frac{\rho}{\varepsilon}\left(\frac{\partial\mathbf{u}}{\partial t} + \frac{1}{\varepsilon}\mathbf{u}\cdot\nabla\mathbf{u}\right) = -\nabla p - \frac{\mu}{K}\mathbf{u} - \frac{\rho F}{\sqrt{K}}\mathbf{u}|\mathbf{u}| + \frac{\mu}{\varepsilon}\nabla^2\mathbf{u} \qquad (2.23)$$

Equation 2.23 is called the *non-Darcy model* of fluid flow in a porous medium and is applicable over a range of Reynolds numbers [5, 8, 9, 16–18].

When gravity effects are important, pressure can be replaced by the effective pressure as in Eq. 2.19. This approach is applicable for all body force fields χ that are expressible as a gradient of a scalar potential ($\chi = -\nabla\phi$). Then, $\tilde{p} = p + \phi$.

Equation 2.23 can be seen from other viewpoints as discussed in the following remarks.

i. The non-Darcy model represents flow in a homogenized porous medium where the building block is the representative elementary volume (REV). Local averaging yields new parameters such as ε, K, and F apart from fluid properties ρ and μ.

ii. The fluid phase in the pores of the physical domain pass through a tortuous path may suffer local adverse pressure gradients and flow separation. These classify as acceleration but are submerged in the volume-averaging step over the REV, leading to the Forschheimer correction. The acceleration terms of Eq. 2.23 will be explicitly nonzero when inflow velocities change with time, the macroscopic flow passage is of variable area, and over geometries such as a cylinder buried in a porous formation.

iii. The basic unknowns of the non-Darcy model are the velocity components that are derivable from the momentum equations and pressure. The latter is obtained from the mass conservation equation, which for an incompressible, fluid saturated porous medium is given by Eq. 2.4 as

$$\nabla \cdot \mathbf{u} = 0$$

iv. Uncertainties in the specification of the effective viscosity (approximated often as the fluid viscosity) and the correlations for K and F contribute to the overall model uncertainty. In turn, these quantities are to be found from benchmark experiments where the flow distribution and pressure drop can be easily measured. The overall approach of extracting the constants of proportionality from experiments falls in the domain of *parameter estimation.*

v. For a homogeneous fluid, $\varepsilon \to 1$ and $K \to \infty$ and Eq. 2.23 reduces to the classical Navier–Stokes equations [11]

$$\rho\left(\frac{\partial \mathbf{u}}{\partial t} + \mathbf{u}\cdot\nabla\mathbf{u}\right) = -\nabla p + \mu\nabla^2\mathbf{u}$$

vi. For definiteness, consider flow in a parallel plate channel filled with a homogeneous and isotropic porous medium. Following the initial and boundary conditions employed in the context of fluid flow problems, those for the non-Darcy model can be stated as follows.

prescribed or quiescent initial flow field,
prescribed inlet velocity distribution,
no-slip and impermeability conditions for velocity at solid surfaces,
fully developed flow at the exit plane, and
pressure specified at one point on the exit plane.

vii. The non-Darcy model shares several mathematical properties with the Navier–Stokes equations. For example, both are second-order nonlinear partial differential equations in space and first order in time. Hence, semi-analytical and numerical methods such as computational fluid dynamics (CFD) applicable for the Navier–Stokes equations can be carried forward to flow in porous media [19].

viii. Equation 2.23 is in coordinate-free form and can be expanded in a coordinate system of interest by using the correct form of the derivative operators. Let (u, v, w) represent the Cartesian components of velocity in the (x, y, z) coordinate system. The three-dimensional unsteady form of the non-Darcy equations can be expressed as follows:

$$\frac{\partial u}{\partial x} + \frac{\partial v}{\partial y} + \frac{\partial w}{\partial z} = 0 \tag{2.24}$$

$$
\begin{aligned}
\frac{\rho}{\varepsilon}\left(\frac{\partial u}{\partial t} + \frac{1}{\varepsilon}\left(u\frac{\partial u}{\partial x} + v\frac{\partial u}{\partial y} + w\frac{\partial u}{\partial z}\right)\right) &= -\frac{\partial p}{\partial x} - \frac{\mu}{K}u \\
&\quad -\frac{\rho F}{\sqrt{K}}u\sqrt{u^2+v^2+w^2} + \mu\left(\frac{\partial^2 u}{\partial x^2} + \frac{\partial^2 u}{\partial y^2} + \frac{\partial^2 u}{\partial z^2}\right) \\
\frac{\rho}{\varepsilon}\left(\frac{\partial v}{\partial t} + \frac{1}{\varepsilon}\left(u\frac{\partial v}{\partial x} + v\frac{\partial v}{\partial y} + w\frac{\partial v}{\partial z}\right)\right) &= -\frac{\partial p}{\partial y} - \frac{\mu}{K}v \\
&\quad -\frac{\rho F}{\sqrt{K}}v\sqrt{u^2+v^2+w^2} + \mu\left(\frac{\partial^2 v}{\partial x^2} + \frac{\partial^2 v}{\partial y^2} + \frac{\partial^2 v}{\partial z^2}\right) \\
\frac{\rho}{\varepsilon}\left(\frac{\partial w}{\partial t} + \frac{1}{\varepsilon}\left(u\frac{\partial w}{\partial x} + v\frac{\partial w}{\partial y} + w\frac{\partial w}{\partial z}\right)\right) &= -\frac{\partial p}{\partial z} - \frac{\mu}{K}w \\
&\quad -\frac{\rho F}{\sqrt{K}}w\sqrt{u^2+v^2+w^2} + \mu\left(\frac{\partial^2 w}{\partial x^2} + \frac{\partial^2 w}{\partial y^2} + \frac{\partial^2 w}{\partial z^2}\right)
\end{aligned}
\tag{2.25}
$$

ix. Once the local flow field and pressure distribution are solved for quantities such
as flow rate and surface forces on a global scale can be obtained by suitably
integrating the local distributions of velocity, pressure, and shear stresses.
Equations 2.24 and 2.25 are coupled in terms of velocity components and
pressure. Coupling, nonlinearity, and the overall complexity of the governing
equations make analysis of the flow field quite daunting.

2.4.1   Non-dimensionalization

Let L be a macroscopic length scale and U a velocity scale chosen for the device or
the apparatus of interest. For homogeneous fluid flow, the scaling law for dimensionless quantities leads to the definition of the Reynolds number

$$\text{Re} = \frac{\rho U L}{\mu}$$

In dimensionless form, velocity is interpreted as the scaled variable u/U and
pressure as $(p - p_{\text{ref}})/\rho U^2$, while the independent variables are scaled as $x \to x/L$
and $t \to Ut/L$. During the process of scale analysis, the continuity equation remains
unchanged.

The dimensionless Navier–Stokes equations written in coordinate-free form can
be readily derived as [11]

$$\left(\frac{\partial \mathbf{u}}{\partial t} + \mathbf{u} \cdot \nabla \mathbf{u}\right) = -\nabla p + \frac{1}{\text{Re}} \nabla^2 \mathbf{u} \tag{2.26}$$

Equation 2.26 shows that two different flow fields with unequal velocities and
dimensions can still be *dynamically similar*, if their Reynolds numbers are equal.
Then, the solution of one in dimensionless form will truly hold for the other.

Extending the scaling analysis further, the non-Darcy equations in
coordinate-free form read as follows:

$$\frac{1}{\varepsilon}\left(\frac{\partial \mathbf{u}}{\partial t} + \frac{1}{\varepsilon}\mathbf{u} \cdot \nabla \mathbf{u}\right) = -\nabla p - \frac{1}{\text{Re} \times Da}\mathbf{u} - \frac{f}{\sqrt{Da}}\mathbf{u}|\mathbf{u}| + \frac{1}{\varepsilon \text{Re}}\nabla^2 \mathbf{u} \tag{2.27}$$

Here, Da, the Darcy number is defined as

$$Da = \frac{K}{L^2} \quad \text{while} f = \frac{1.8}{(180\varepsilon^5)^{0.5}}\varepsilon$$

Apart from Reynolds number, ε, Da, and f appear as additional dimensionless
parameters in the formulation. For a given Reynolds number, the flow details will

then depend primarily on porosity, permeability, and the inertia coefficient; correspondingly, the flow distribution will depend on the dimensionless parameters of the formulation including ε, Da, and f.

2.4.2 Special Cases

The non-Darcy model is quite complex in its general form, but can be simplified in many circumstances. For steady flow, the dependence on time can be dropped, namely

$$\frac{\partial \mathbf{u}}{\partial t} = 0$$

When the flow is fully developed, the acceleration terms can be set to zero; hence

$$\left(\frac{\partial \mathbf{u}}{\partial t} + \frac{1}{\varepsilon} \mathbf{u} \cdot \nabla \mathbf{u} \right) = 0$$

Accordingly, one solves the following system of equations for the flow field:

$$0 = -\nabla p - \frac{1}{\text{Re} \times Da} \mathbf{u} - \frac{f}{\sqrt{Da}} \mathbf{u} |\mathbf{u}| + \frac{1}{\varepsilon \text{Re}} \nabla^2 \mathbf{u}$$

Additional simplifications may arise from the reduced dimensionality of the flow field. It is also clear that for small Reynolds numbers, the square-law dependence on velocity can be dropped; hence

$$\text{Re} < < 1, \quad -\nabla p - \frac{1}{\text{Re} \times Da} \mathbf{u} + \frac{1}{\varepsilon \text{Re}} \nabla^2 \mathbf{u} = 0$$

Equating the second and the third terms, the boundary-layer thickness δ near a solid wall can be estimated as

$$\frac{\delta}{L} \sim \sqrt{\frac{Da}{\varepsilon}}$$

Hence, in low permeability porous media $(Da < < 1)$, the boundary-layer thickness is small and the corresponding wall shear stress term can be omitted in the calculation of the flow field outside the boundary layer. Thus, under the twin approximations of low Reynolds number and low permeability, we recover Darcy's law as the valid form of the momentum equation:

$$\text{Re} < < 1, Da < < 1 : -\nabla p - \frac{1}{\text{Re} \times Da}\mathbf{u} = 0$$

The above equation will be adequate for pressure drop calculations but will not resolve the wall boundary layers.

2.4.3 Compressible Flow

Compressible flow of a liquid in a Darcian framework is discussed in Sect. 2.1.3 in terms of a fluid property, namely compressibility. The present discussion generalizes this approach for a non-Darcy model.

Compressibility of the fluid in the porous region introduces fundamental changes in the mathematical equations relative to the incompressible formulation. Specifically, the mass balance equation is used to determine fluid density while the equation of state determines pressure. For a gas, changes in pressure take place simultaneously with temperature that must be introduced as a dependent variable. Thus, mass, momentum, and energy equations are coupled and are to be simultaneously solved. While working with compressible flow, it is common to introduce auxiliary-dependent variables such as enthalpy and entropy, depending on the application of interest. For example, when shocks are formed, large changes in entropy take place and it is advantageous to introduce entropy as a dependent variable.

The present discussion is restricted to gas flow in a fixed porosity porous region where compressibility arises from high gas speeds and hence, large changes in pressure. The fluid medium is taken to be an ideal gas so that the following equation of state, connecting pressure, density, and temperature, is applicable:

$$\frac{p}{\rho} = RT \tag{2.28}$$

In view of large changes in specific heat $C_p(T)$ of the gas, it is preferable to work with enthalpy h that relates to temperature as

$$C_p(T) = \frac{\partial h}{\partial T}\bigg|_p$$

For high speed flow of compressible gas in a homogeneous region with no chemical reactions and shaft work, the governing equations are of mass, momentum, and energy (first law of thermodynamics), stated as follows [20]:

$$\frac{\partial \rho}{\partial t} + \nabla \cdot \rho \mathbf{u} = 0$$

$$\rho \left(\frac{\partial \mathbf{u}}{\partial t} + \mathbf{u} \cdot \nabla \mathbf{u} \right) = -\nabla p + \nabla \cdot \left(\mu (\nabla \mathbf{u} + \nabla \mathbf{u}^T) - \frac{2}{3} \delta_{ij} \nabla \cdot \mathbf{u} \right) \qquad (2.29)$$

$$\rho \left(\frac{\partial e}{\partial t} + \mathbf{u} \cdot \nabla e \right) = \nabla \cdot \left(\frac{k}{C_p} \right) \nabla e \qquad (e = h + \frac{\mathbf{u}^2}{2})$$

In this approach, the momentum equations solve for velocity, mass balance for density, energy equation for enthalpy and temperature while pressure is determined form the equation of state (2.28). Nevertheless, the system of equations is tightly coupled and must be jointly solved for the dependent variables. Since compressibility effects arise jointly with high speed flow of the gaseous medium, inertia effects in the momentum equation and advection terms in the energy equation become prominent. Equation 2.29 can then be simplified by dropping the viscous terms in the momentum equation and diffusion terms in the energy equations. These approximations considerably reduce the effort required to solve the governing equations.

Energy equation for a porous region is discussed in Sect. 2.5. The compressible flow equations as applied for a porous medium are subsequently revisited in Sect. 2.5.2.

2.4.4 Turbulent Flow

The overall effect of the solid grains of a porous matrix is to stabilize flow by increasing viscous forces uniformly throughout the physical domain. Conversely, when the pore size becomes large, the pore-scale Reynolds number can be substantial, leading to persistent unsteadiness in the local flow field. Boundary-layer thickness near an impermeable wall is small in a dense porous medium, scaling with $\sqrt{K}$ (K being the permeability). This change in boundary-layer thickness suggests a delay in the transition to turbulence [21]. To a first approximation, it can now be assumed that large-scale vortices of the size of the boundary-layer thickness do not appear in the porous medium. It is likely that the main sources of turbulence are then (i) vortex formation from the surface of the solid particles and (ii) shear layer instability in the pore space. The respective length scales of the turbulent fluctuations are then the particle diameter and the pore diameter [22, 23]. With the size of the REV much greater than either of these scales, the effect of fluctuations on flow distribution and pressure drop will have to be factored in through the time domain statistics.

While resolving every timescale present in the time-dependent fluctuations of velocity can be numerically difficult, a Reynolds-averaging approach can be attempted for turbulent flow. This step follows volume-averaging over REV that is intrinsic to the mathematical modeling of flow in porous media [24]. The time-averaged flow field is now modified by Reynolds stresses that are temporal correlations of the velocity fluctuation vector. Combined with the Boussinesq approximation of turbulent viscosity, one obtains a modified form of the momentum equation wherein fluid viscosity is replaced by the sum of fluid viscosity and an eddy (or, turbulent) viscosity.

The time-averaged approach to modeling of turbulent flow ensures that the overall (time-averaged) pressure drop is correctly predicted for a given flow rate. In the process, instantaneous changes in velocity and pressure, particularly in the pores, are not fully resolved. The non-Darcy model accounts for the Darcian pressure drop arising from the interfacial friction between the fluid and solid phases, viscous friction within the fluid medium via the Brinkman correction, and flow separation effects including vortex shedding through the Forschheimer terms. Additional corrections for the turbulent fluctuations will necessarily represent those at the small scales within the pores of the porous medium. These effects, particularly vortex formation and turbulent fluctuations, are overlapping. As a result, modeling parameters of turbulent flow in porous media will be distinct from a homogeneous fluid and must be individually obtained from dedicated experiments of parameter estimation.

Following Eq. 2.27, the momentum equation governing turbulent flow in a porous medium can be written as

$$\frac{1}{\varepsilon}\left(\frac{\partial \mathbf{u}}{\partial t} + \frac{1}{\varepsilon}\mathbf{u}\cdot\nabla\mathbf{u}\right) = -\nabla p - \frac{1}{\mathrm{Re}\times Da}\mathbf{u} - \frac{f}{\sqrt{Da}}\mathbf{u}|\mathbf{u}|$$
$$+ \frac{1}{\varepsilon\mathrm{Re}}\nabla\cdot\left(1 + \frac{\mu_T}{\mu}\right)\nabla\mathbf{u}$$

Here, velocities and pressure are both REV- and time-averaged. Symbol μ_T represents turbulent viscosity and is zero below a certain threshold Reynolds number. The continuity equation in terms of the time-averaged velocity will be applicable for mass conservation. Following the discussion on dispersion of Sect. 2.5, turbulent viscosity can be obtained as

$$\frac{\mu_T}{\mu} = C\mathrm{Re}_l, \mathrm{Re}_l > 100 \text{ and } 0, \text{ otherwise}$$

Here, Reynolds number is based on a pore-scale dimension and the average fluid speed. The constant of proportionality is expected to be of the order of unity. Here, turbulent fluctuations are taken to be present at the pore scale and create viscosity-like effects seen in jets and wakes. Correlations of turbulent viscosity similar to the one above can be seen in the context of round jets [25, 26]. Higher-order turbulence models (such as $k - \varepsilon$) in porous media have been discussed in the literature [27].

In heat and mass transfer, a non-participating solid phase merely creates a tortuous flow path. Combined with velocity fluctuations, the net influence on transport is to increase dispersion. See discussion accompanying Eq. 2.33 in Sect. 2.5.

Consider a thermally participating medium where the solid phase thermal conductivity is much greater that of the fluid phase. Thermal fluctuations are then uniformly damped within the porous medium (Eq. 2.33 of Sect. 2.5). The effect of turbulence on thermal transport will then be a function of the solid-to-fluid thermal conductivity ratio.

2.5 Energy Equation

Convective heat transfer in a porous region is the subject of discussion here. For a homogeneous fluid, the equation governing heat transfer is derived from the first law of thermodynamics [28]. The following assumptions greatly simplify the governing equation for temperature. These are:

  i. Incompressible flow
 ii. Constant fluid properties that are independent of temperature
iii. Negligible viscous dissipation
 iv. No shaft work (or devices)
  v. No heat generation
 vi. No radiative fluxes

For a homogeneous fluid region, the governing equation is expressed as [29]

$$\rho C \left(\frac{\partial T}{\partial t} + \mathbf{u} \cdot \nabla T \right) = k \nabla^2 T$$

Here, ρ is fluid density, C is specific heat, and k is thermal conductivity. This form of the energy equation can be extended to a porous medium carrying fluid flow. The additional approximations needed to set up the energy equation in a homogeneous isotropic porous region are as follows.

1. The symbol $T(x, y, z, t)$ represents the REV-averaged temperature of the fluid and the solid phases jointly. Thus, the phases are at equal temperatures within the REV, for all time instants, an assumption referred as one of *thermal equilibrium*.
2. The time derivative refers to the change in sensible energy of the porous medium within the REV. Hence, the multiplying factor is $(\rho C)_m$ determined for the porous medium.
3. The advection term refers to energy transfer by the fluid phase. Hence, the multiplier is $(\rho C)_f$ as appropriate for the fluid phase.
4. On the right-hand side, thermal conductivity will be the effective medium conductivity (also called, *the stagnant medium* conductivity), denoted as k_m.
5. Heat conduction in a porous medium continues to be given Fourier's law

$$\mathbf{q} = -k_m \nabla T$$

Thus, the thermal energy equation in the porous medium can be expressed as

$$(\rho C)_m \frac{\partial T}{\partial t} + (\rho C)_f \mathbf{u} \cdot \nabla T = k_m \nabla^2 T \tag{2.30}$$

Expressions relating thermophysical properties of the solid and the fluid phases with those of the medium have been extensively discussed in the literature [9, 15, 30]. These depend on the individual properties, porosity, and the pore structure. The simplest correlations for specific heat and thermal conductivity of the porous region are given by the rule of mixtures:

$$(\rho C)_m = \varepsilon(\rho C)_f + (1 - \varepsilon)(\rho C)_s$$
$$k_m = \varepsilon k_f + (1 - \varepsilon)k_s$$

Here, f and s refer to the fluid and the solid phases while m is the porous medium. Effective (equivalent) thermal conductivity of the porous medium is related to the direction of the temperature gradient, and the expression above is to be seen as an upper limit when the solid and the fluid phases are in series. A parallel arrangement of the phases yields the lower limit as

$$\frac{k_m}{k_f} = \frac{k_s/k_f}{(1 - \varepsilon) + \varepsilon k_s/k_f}$$

The effective thermal conductivity of a sufficiently dilute suspension ($\varepsilon \to 1$) with spherical particles is given by the analytically derived Maxwell model:

$$\frac{k_m}{k_f} = \frac{(3 - 2\varepsilon)k_s/k_f + 2\varepsilon}{3 - \varepsilon + \varepsilon k_s/k_f}$$

An equivalent expression with the solid phase treated as a collection of cylinders (wires) is given by the Rayleigh model:

$$\frac{k_m}{k_f} = \frac{1 + k_s/k_f - (1 - \varepsilon)(1 - k_s/k_f)}{1 + k_s/k_f + (1 - \varepsilon)(1 - k_s/k_f)}$$

Other models for non-periodic geometries and non-spherical particles are presented in the literature, but the effective medium conductivity is a source of uncertainty in transport modeling in porous media. In general, these can be derived mainly from experiments. In a gas–metal system (as in Chap. 6), the medium properties are closely aligned with the metallic phase with $(1 - \varepsilon)$, namely (1—porosity) as a multiplier. For temperature-dependent thermal conductivity, the right side of Eq. 2.30 is written as $\nabla \cdot (k_m(T)\nabla T)$.

For anisotropic and inhomogeneous porous media, thermal conductivity is a second-order tensor (like permeability in Sect. 2.1.2), and the energy equation is written as

$$(\rho C)_m \frac{\partial T}{\partial t} + (\rho C)_f \mathbf{u} \cdot \nabla T = \nabla \cdot [k_m] \nabla T \qquad (2.31)$$

Fluid velocity in these expressions continues to be the Darcian velocity evaluated for the REV as a whole. At the scale of the REV, the real flow path tends to be tortuous, thus increasing the possibility of intermixing of the hot and the cold fluid streams and greater spreading. This effect will show up as an augmentation of thermal conductivity and is referred as *dispersion* (Fig. 2.3). Since velocity has three components, dispersion can occur in three dimensions. Equation 2.31 is now written to include a dispersion tensor, also called dispersion coefficient D_m as

$$(\rho C)_m \frac{\partial T}{\partial t} + (\rho C)_f \mathbf{u} \cdot \nabla T = \nabla \cdot [k_m + D_m] \nabla T \qquad (2.32)$$

Here, dispersion augments molecular conduction and so the elements of the dispersion tensor are positive. Experiments in homogeneous isotropic porous media show that dispersion tensor reduces to a scalar quantity and is related to fluid speed as [31, 32]:

$$d_m = (\rho C)_m l |\mathbf{u}| \text{ where}$$
$$l = \text{constant} \cdot d_p (Pe_d < 1); l = C Pe_d^n \cdot d_p (Pe_d > 1), 0 < n < 1$$

Symbol l is called *dispersion length*, and d_p is the particle diameter. Peclet number is defined in this context as

$$Pe_d = \frac{|\mathbf{u}| d_p}{\alpha_f}$$

Here, α_f is the fluid thermal diffusivity.

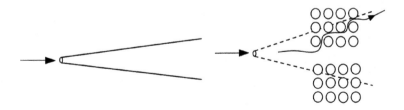

Fig. 2.3 Schematic drawing of tortuosity and dispersion in a porous medium; (left) spread of thermal energy from a line source in a homogeneous fluid phase; (right) higher spread in a porous medium arising from tortuosity of the flow path

Let U be the velocity scale, L the length scale, and the property ratios defined as

$$\tilde{\beta} = \frac{(\rho C)_f}{(\rho C)_m} \quad \lambda = \frac{k_f}{k_s}$$

The dimensionless form of the thermal energy equation with constant dispersion is written as

$$\frac{\partial T}{\partial t} + \tilde{\beta}\mathbf{u} \cdot \nabla T = \nabla \cdot \left(\frac{1}{Pe}\nabla T\right) \tag{2.33}$$

Peclet number appropriate for the scale of the entire porous medium is now defined as

$$Pe = \frac{UL}{\alpha_{mol} + \alpha_m}$$

The molecular thermal diffusivity in the definition of Pe is

$$\alpha_{mol} = \frac{k_m}{(\rho C)_m}$$

while the dispersion part is

$$\alpha_m = \frac{d_m}{(\rho C)_m} = l|\mathbf{u}| \sim lU$$

In the presence of turbulent fluctuations in the pores of the physical region, the dispersion length is further augmented by an amount, constant × pore diameter (Sect. 2.4.4), where the constant value is to be determined from reference experiments. Being dimensionless, the constant will scale with the fluid Prandtl number. It is understandable that velocity and temperature fluctuations will correlate for small Prandtl numbers as in liquid metals. For $Pr \geq 1$, this dependence can be expected to be small [25]. For small fluid-to-solid thermal conductivity ratio λ, thermal fluctuations will be damped by the solid phase.

The thermal energy equation requires specification of additional parameters over and above the flow field itself. Being a second-order partial differential equation in space, boundary conditions must also be specified. As in the modeling of pressure and velocity, these are classified as Dirichlet (temperature specified), Neumann (heat flux specified), and Robin (linear combination of temperature and heat flux specified). For unsteady heat transfer problems, an initial condition is required for the temperature calculation within the porous medium.

2.5.1   Thermal Non-equilibrium Model

The assumption of equal phase temperatures is equivalent to requiring the fluid and the solid media attain local thermal equilibrium quickly, in relation to the overall timescale of the transport process. This assumption is violated in several contexts and more so during rapid transients, namely when the timescale of the superimposed process is itself small.

An explicit treatment of individual phase temperatures is more general and will contain the approximation of thermal equilibrium as a special case. In view of this generality, the thermal non-equilibrium model is presented below. It is assumed that the porous medium has a solid phase coexisting with a single continuous fluid phase, leading to the two-equation model for temperature. These equations solve for the fluid and solid phase temperatures.

The respective equations will resemble the energy equations at the pure fluid and the pure solid limits with a point of difference. For the fluid medium, the solid particles at a distinct temperature will serve as an energy *sink* (or *source*, depending on the sign of the local temperature difference). One can imagine a hot fluid entering a cold porous region; a substantial part of the fluid energy will be spent in warming up the solid phase. This process requires a temperature difference and, clearly, a non-equilibrium model. The fluid–solid energy exchange term is thus central to the two-equation formalism.

The fluid phase energy equation for temperature T^f is written as:

$$\varepsilon(\rho C)_f \left(\frac{\partial T^f}{\partial t} + \frac{\mathbf{u}}{\varepsilon} \cdot \nabla T^f \right) = k_f \nabla^2 T^f + Q_{f \leftarrow s} \tag{2.34}$$

Similarly, the energy equation for the solid phase is written in terms of temperature T^s as

$$(1 - \varepsilon)(\rho C)_s \left(\frac{\partial T^s}{\partial t} \right) = k_s \nabla^2 T^s + Q_{f \rightarrow s} \tag{2.35}$$

Comparing Eqs. 2.34 and 2.35, the following points of difference may be noted:

(a) The heat capacity term for the fluid is multiplied by porosity which is the volume fraction of the fluid phase. The corresponding term in the solid phase equation is multiplied by (1-porosity), the volume faction of the solid phase. This is understandable because the two equations are jointly applicable for a unit volume of the porous medium.

(b) The fluid temperature equation is written exclusively in terms of its thermophysical properties. The solid phase equation employs its own conductivity, density, and specific heat. It is assumed here that, within the REV, the two phases are adequately connected with independent paths for heat transfer.

(c) Since energy exchange at the phase boundary is continuous at all time instants, the source terms $Q_{f \to s}$ and $Q_{f \leftarrow s}$ are equal in magnitude but opposite in sign.

Equations 2.34 and 2.35 can be used in modeling only after the source terms are prescribed. Assuming a convective heat transfer environment, the source term, evaluated per unit volume of the REV, can be written as

$$Q_{f \to s} = hA_{IF}(T^f - T^s) \tag{2.36}$$

Similarly

$$Q_{f \leftarrow s} = hA_{IF}(T^s - T^f)$$

The heat transfer coefficient between the two phases is denoted as h and A_{IF}, the *specific area*, is the interfacial area per unit volume of the porous medium (units of m^{-1}). The specific area will depend on the pattern of arrangement of the solid phase within the fluid. Heat transfer coefficient can be determined from Nusselt number correlations such as

$$Nu = \frac{hL}{k_f} = APe^m$$

The diffusion term in Eq. 2.34 can be rewritten to include dispersion as

$$\nabla \cdot (k_f + d_f)\nabla T = \nabla \cdot (k_{eff})\nabla T$$

Using uniform velocity and length scales U and L for the two temperature equations, a timescale of L/U, and $Pe = UL/\alpha_f$, the dimensionless form of the thermal non-equilibrium model is summarized as follows:

$$\text{FLUID:}\quad \varepsilon\left(\frac{\partial T^f}{\partial t} + \frac{\mathbf{u}}{\varepsilon} \cdot \nabla T^f\right) = \frac{1}{Pe}\nabla \cdot \left(\frac{k_{eff}}{k_f}\right)\nabla T^f$$
$$- \frac{NuA_{IF}}{Pe}(T^f - T^s) \tag{2.37}$$

$$\text{SOLID:}\ (1-\varepsilon)\left(\frac{\partial T^s}{\partial t}\right) = \frac{\beta/\lambda}{Pe}\nabla^2 T^s$$
$$+ \frac{NuA_{IF}\beta}{Pe}(T^f - T^s) \quad \text{where } \beta = \frac{(\rho C)_f}{(\rho C)_s} \tag{2.38}$$

Symbol λ is the fluid-to-solid thermal conductivity ratio. An application of the two-equation model is presented in Chap. 6 for oscillatory flow in metallic mesh regenerators.

The two-equation model of heat transfer will reduce the thermal equilibrium model when the energy exchange term of Eq. 2.36 is zero. This limit is reached when the heat transfer coefficient is large, indicating $T^f - T^s \to 0$, i.e., an equality of phase temperatures. Alternatively, the solid phase conductivity can approach zero, thus decoupling the solid and fluid phase temperatures. In this respect, the thermal equilibrium model also holds for a thermally non-participating solid phase whose role is entirely to create a tortuous flow path for the fluid medium.

Effective thermal conductivity in Eq. 2.37 will have to be augmented for additional dispersion in turbulent flow. Temperature fluctuations will be damped by a highly conducting solid phase, and this correction will not be required. In a solid phase of low conductivity, a single-equation model of heat transfer will be adequate. Here, the correction for turbulent fluctuations will be as per the discussion accompanying Eq. 2.33 (Sect. 2.5).

2.5.2 Compressible Flow

Following the discussion of Sect. 2.4.3, we examine the mass, momentum, and energy equations of high speed gas flow in a porous medium whose pore geometry as well as the solid particles is rigid. Mechanical work and viscous dissipation terms are neglected in the fluid phase energy equation with the expectation that the fluid–solid energy exchange terms would be dominant. Total energy, the sum of enthalpy and kinetic energy is approximated as enthalpy since velocities expected within a porous medium will not be excessive. Enthalpy is written as the product of (ρC) and T, for each of the two phases. A non-Darcy, two-equation thermal non-equilibrium model is proposed here, and the following system of equations can be derived:

$$
\text{Mass}: \varepsilon \frac{\partial \rho}{\partial t} + \nabla \cdot (\rho \mathbf{u}) = 0
$$

$$
\text{Momentum}: \frac{\rho}{\varepsilon}\left(\frac{\partial \mathbf{u}}{\partial t} + \frac{1}{\varepsilon}\mathbf{u} \cdot \nabla \mathbf{u}\right) = -\nabla p - \frac{\mu}{K}\mathbf{u} - \frac{\rho F}{\sqrt{K}}\mathbf{u}|\mathbf{u}| + \frac{\mu}{\varepsilon}\nabla^2 \mathbf{u}
$$

$$\text{(2.39)}$$

$$
\text{Fluid}: \varepsilon\left(\frac{\partial (\rho C)_f T^f}{\partial t} + \nabla \cdot (\rho C)_f \frac{\mathbf{u}}{\varepsilon} T^f\right) = \nabla \cdot \left(\frac{k_{eff\,f}}{C_f}\nabla(C_f T^f)\right) + Q_{f \leftarrow s}
$$

$$
\text{Solid}: (1 - \varepsilon)\left(\frac{\partial (\rho C)_s T^s}{\partial t}\right) = \nabla \cdot \left(\frac{k_s}{C_s}\nabla(C_s T^s)\right) + Q_{f \to s}
$$

The fluid–solid energy exchange term is given by Eq. 2.36. The left side of the fluid phase energy equation can be simplified using the mass balance equation as

$$
\varepsilon\left(\frac{\partial (\rho C)_f T^f}{\partial t} + \nabla \cdot (\rho C)_f \frac{\mathbf{u}}{\varepsilon} T^f\right) = \varepsilon \rho_f \left(\frac{\partial (C_f T^f)}{\partial t} + \frac{\mathbf{u}}{\varepsilon} \cdot \nabla(C_f T^f)\right)
$$

The form on the right side is convenient for discretization in numerical simulations. The product of specific heat and temperature can be treated as a single dependent variable. The temperature formulation is, however, to be preferred since the energy exchange term is presented in the form the phase temperature difference. Specific heat as a function of temperature for the participating phases is independently supplied as constitutive equations in the model. Thus, in the system given by Eq. 2.39, the mass balance equation determines density, momentum equation solves for the Darcian velocity, the energy equations yield the fluid and solid phase temperatures, and the ideal gas equation of state will determine the fluid phase pressure. For a discussion on compressible flow through a deformable porous medium and formation of shock waves, see [33, 34].

2.6 Unsaturated Porous Medium

In the discussion of Sects. 2.1–2.5, the pore space was taken to be filled with a single fluid, thus making the porous medium saturated. Consider flow of water in a porous medium with air pockets around. An example is groundwater flow, where water may not fill the available pore space. Such a medium is said to be *unsaturated*. Applications involving unsaturated flow are quite a few. A variable that characterizes the volume fraction of the liquid in the pore space available within the porous medium is *saturation* and is denoted as S. In the context of groundwater, the symbol adopted is S_w to indicate water saturation. The volume fraction of air in such an instance is $1 - S_w$.

An important consequence of the porous medium being unsaturated is the appearance of interfacial forces. Figure 2.4 shows a porous medium partially saturated with water. Assuming that the solid phase is wetted by water, a meniscus will form within the pore space, as shown schematically in Fig. 2.4. A simple balance of forces at the interface shows that, under equilibrium conditions, pressure in the liquid phase will be smaller than that in the gaseous phase, namely $p_w < p_a$..

Fig. 2.4 Formation of interfaces and the importance of surface tension in an unsaturated porous medium containing air and water. As shown, water wets the solid surface and is called the *wetting* phase

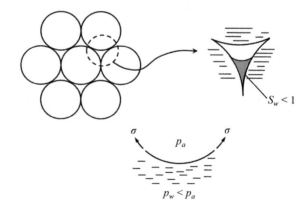

The difference is balanced by the appropriate component of surface tension integrated over the wetting perimeter. The difference will scale with the interface curvature, increasing as the pore size diminishes and decreasing with higher water content in the pores.

Suppose the quantity of interest is flow of water, for example, the speed with which it saturates the porous medium. To a first approximation, we may assume here that the gaseous phase, a mixture of air and moisture, will be practically at the atmospheric pressure. To mobilize water, an extra pressure will have to be provided at one end of the porous region. Flow is setup when the excess pressure overcomes the force of surface tension at the water–air–grain three-phase boundary.

An exact analysis of interfacial forces in a porous medium is formidable, and more so because it depends on the geometry of the pore space. Instead, analysis can proceed on the scale of the REV, and one can write the revised form of Darcy's law, applicable for a phase ϕ of an unsaturated porous medium as follows [35, 36]:

$$\mathbf{u}^\phi = -k_r(S_\phi)\frac{K}{\mu}\nabla p^\phi \tag{2.40}$$

Comparing with Eq. 2.1, it is clear that the difference in formulation arises from the function k_r, called the *relative permeability*. It is a dimensionless quantity. Since the formation of interfaces increases flow resistance, it can be surmised that the relative permeability function will be smaller than unity. When water is withdrawn from the pores, the interface curvature steepens, increasing the flow resistance further. Thus, relative permeability is a function of the phase saturation, a functionality shown in Eq. 2.40.

In a porous medium with two fluids 1 and 2 jointly present, Eq. 2.40 is written as follows:

$$\begin{aligned}
\mathbf{u}_1 &= -k_{r1}(S_1)\frac{K}{\mu_1}\nabla p_1 \\
\mathbf{u}_2 &= -k_{r2}(S_2)\frac{K}{\mu_2}\nabla p_2
\end{aligned} \tag{2.41}$$

Here, $S_1 + S_2 = 1$. For air–water flow, air may be taken to be at the atmospheric pressure, namely $p_2 \approx p_{atm}$ leading to zero velocity in the gaseous phase. Equation 2.41 then reduces to a single momentum equation for the migration of water:

$$\mathbf{u}_w = -k_r(S_w)\frac{K}{\mu_w}\nabla p_w$$

Equations 2.40 and 2.41 are momentum equations for unsaturated flow in porous media. The restrictions applicable for Darcy's law continue to hold, namely the pore-scale Reynolds number is less than unity. This approximation is quite acceptable in many instances since fluid speeds are smaller under unsaturated conditions.

The phase pressures are, in turn, determined with reference to the mass balance equation for an unsaturated medium. The discussion is initially for air–water flow where air is the stagnant phase. Water saturation is denoted as S_w, ρ is water density, and $\mathbf{u}$ is the Darcian velocity of water averaged over the REV. Start with the general form of the mass balance equation for the liquid contained in the pores of the physical region. Hence

$$\frac{\partial}{\partial t}(\varepsilon \rho S_w) + \nabla \cdot \rho \mathbf{u} = 0$$

The product of porosity and water saturation arises from the fact that saturation is defined with respect to the pore volume. Treating water as incompressible (ρ = constant) and porosity as constant, the mass balance equation simplifies to

$$\varepsilon \frac{\partial S_w}{\partial t} + \nabla \cdot \mathbf{u} = 0 \tag{2.42}$$

Using Eq. 2.41 for velocity, the pressure equation can be derived as

$$\varepsilon \frac{\partial S_w}{\partial t} = \frac{\partial}{\partial x}\left(\frac{k_r(S_w)K}{\mu_w}\frac{\partial p_w}{\partial x}\right) \tag{2.43}$$

Permeability K and water viscosity μ_w are independently prescribed quantities. Relative permeability is a function of water saturation and is a characteristic property of the porous medium. It is to be specified from experimental characterization of the medium under study. Since water migration in unsaturated soils is a slow process, quasi-static conditions may be assumed to hold. The pressure difference across the interface between the non-wetting and the wetting phases, called the *capillary pressure* (p_c), will be function of the interface shape and liquid volume. To a first approximation, this dependence can be linked with the local saturation [37]. Thus, for air–water flow (Fig. 2.4), capillary pressure, a positive quantity, is given as

$$p_c = p_a - p_w = g(S_w)$$

Typical variations of relative permeability with saturation and saturation with capillary pressure are shown in Fig. 2.5.

Equation 2.43 can be written entirely in the form of a second-order nonlinear PDE for water pressure as follows:

$$-\varepsilon \frac{dS_w}{dp_c}\frac{\partial p_w}{\partial t} = \frac{\partial}{\partial x}\left(\frac{k_r(S_w)K}{\mu_w}\frac{\partial p_w}{\partial x}\right) \tag{2.44}$$

Equation 2.44 will determine phase pressure in water, and fluid velocity can be determined from Darcy's law (Eq. 2.41). The derivative of water saturation with

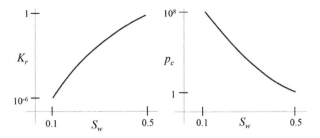

Fig. 2.5 Variation of relative permeability with water saturation and saturation with capillary pressure (arbitrary units) in clay-type porous media

capillary pressure is negative (see Fig. 2.5). With this comment, Eq. 2.44 is in the form of a nonlinear diffusion equation [38].

Under non-isothermal conditions, it is possible to identify distinct temperatures for the fluid phases and one for the solid phase. In a thermal equilibrium model, the three are assigned a single volume-averaged temperature and Eq. 2.32 carries over. Effective medium properties of thermal capacity and thermal conductivity are now functions of the local saturation.

Extensions of the formulation for unsaturated flow with two or more fluids are discussed in the following sections.

2.6.1 Oil–Water Flow

As an immediate extension of unsaturated flow of water in a partially wet porous region, immiscible flow of oil and water in a porous medium is now considered. The context is an application of enhanced oil recovery from oil-rich reservoirs. Pressurized water serves to mobilize oil trapped in the pores of the host rock. Heating the injected water also serves to lower oil viscosity and decrease flow resistance. The physical picture outlined in the previous section carries over. Oil and water form an interface, and surface tension creates a jump in pressure across this boundary. Thus, the two fluid phases move under the influence of the respective phase pressures. Unlike air that was taken to be at the atmospheric pressure, water and oil are in motion, leading to a spatiotemporal variation in the fluid pressures. The overall approach of using relative permeability to augment flow resistance and deriving capillary pressure from saturation continues to be applicable. As expected, these equations are nonlinear, and specialized numerical methods are needed for their solution. The subject of oil–water flow has been extensively studied over the years, and several references can be seen in the list at the end of the chapter [39–42].

Mass balance: The mass balance equation is written for oil and water as follows:

$$\text{Oil}: \frac{\partial}{\partial t}(\varepsilon \rho_o S_o) + \nabla \cdot (\rho_o \mathbf{u}_o) = 0 \tag{2.45a}$$

$$\text{Water}: \frac{\partial}{\partial t}(\varepsilon \rho_w S_w) + \nabla \cdot (\rho_w \mathbf{u}_w) = 0 \tag{2.45b}$$

Specifying that that the pores contain water and oil and no other fluid, we get

$$S_o + S_w = 1 \tag{2.46}$$

Momentum equation: Assuming the fluid velocities to be reasonably small, the momentum equation is a form of the generalized Darcy's law that includes a correction by the relative permeability. For oil and water, these are expressed as follows:

$$\text{Oil}: \mathbf{u}_o = -\frac{k_{ro}(S_o)K}{\mu_o} \nabla p_o \tag{2.47a}$$

$$\text{Water}: \mathbf{u}_w = -\frac{k_{rw}(S_w)K}{\mu_w} \nabla p_w \tag{2.47b}$$

The momentum equations can be combined with the mass balance equations to arrive at a single nonlinear second-order partial differential equation in pressure for each of the two phases. Here, the temperature field (REV-averaged) is taken to be evolving in space and time so that the fluid properties—density and viscosity—are non-constant. Fluid compressibility with respect to pressure and temperature is defined as:

$$\xi = \left.\frac{\partial \rho}{\partial p}\right|_T ; \qquad \zeta = \left.\frac{\partial \rho}{\partial T}\right|_p$$

The phase pressure equations can now be derived as:

$$\text{Oil}: -S_o \zeta_o \frac{\partial T}{\partial t} + S_o \xi_o \frac{\partial p_o}{\partial t} - \frac{dS_w}{dp_c}\left[\frac{\partial p_o}{\partial t} - \frac{\partial p_w}{\partial t}\right]$$
$$= \frac{1}{\rho_o} \nabla \cdot \left[\frac{K k_{ro} \rho_o}{\varepsilon \mu_o}\right] \nabla p_o \tag{2.48a}$$

$$\text{Water}: -S_w \zeta_w \frac{\partial T}{\partial t} + S_w \xi_w \frac{\partial p_w}{\partial t} - \frac{dS_w}{dp_c}\left[\frac{\partial p_o}{\partial t} - \frac{\partial p_w}{\partial t}\right]$$
$$= \frac{1}{\rho_w} \nabla \cdot \left[\frac{K k_{rw} \rho_w}{\varepsilon \mu_w}\right] \nabla p_w \tag{2.48b}$$

Here, water is the wetting phase with oil, the non-wetting phase. Consequently, the capillary pressure, a positive quantity, is expressed as $p_c = p_o - p_w$ and $S_o + S_w = 1$. In an isothermal reservoir, derivatives of temperature can be set equal to zero.

Thermal energy equation: If the injected water is at a temperature higher than the oil-rich formation, a thermal front will move through the porous region. The front position can be determined by solving the appropriate form of the thermal energy equation. For the present discussion, an REV-averaged, thermal equilibrium model is proposed, wherein the fluid phases as well as the porous matrix are locally at the same temperature. This assumption is realistic because the fluid speeds in the reservoir are expected to be small, yielding time constants much larger than the fluid–solid equilibration time. Following Eq. 2.32, the thermal energy equation is written in Cartesian coordinates as:

$$\frac{\partial T}{\partial t} + U\frac{\partial T}{\partial x} + V\frac{\partial T}{\partial y} = \frac{K_h}{\sigma_T}\nabla^2 T \tag{2.49}$$

Here, the coefficients include the effective medium properties corrected for fluid saturation. These are respectively given as follows:

$$
\begin{aligned}
U &= \frac{u_o \rho_o C_o + u_o \rho_o C_o}{\sigma_T} \\
V &= \frac{v_o \rho_o C_o + v_o \rho_o C_o}{\sigma_T} \\
\sigma_T &= \varepsilon(S_o \rho_o C_o + S_w \rho_w C_w) + (1-\varepsilon)(\rho C)_R
\end{aligned}
\tag{2.50}
$$

Symbol K_h represents the effective thermal conductivity of the porous medium filled with multiple fluids. As written in Eq. 2.49, dispersion is not considered since fluid speeds are small. If the variation in conductivity with saturation is included, thermal conductivity will no longer be spatially constant (see discussion in Sect. 2.5). The momentum equation is affected by temperature distribution, first through the explicit temperature terms, but this is usually a smaller influence. The real dependence emerges from large reduction in oil viscosity with increasing temperature.

The mass, momentum, and energy equations are to be supplemented with thermophysical properties of oil and water as well as the oil-bearing rock (marked R). In addition, the following constitutive relations specific to the site being analyzed are required:

$$
\begin{aligned}
k_{rw} &= k_{rw}(S_w) \\
k_{ro} &= k_{ro}(S_w) \\
p_c &= p_o - p_w = p_c(S_w) \\
\frac{\rho_i}{\rho_{ref}} &= 1 + \xi_i(p_i - p_{ref}) - \zeta_i(T_i - T_{ref}), i = \text{oil,water}
\end{aligned}
\tag{2.51}
$$

Apart from the variability of porosity and permeability in a field-scale oil-bearing reservoir, parametric functions such as relative permeability and capillary pressure can also vary substantially and over several orders of magnitude (Fig. 2.5). Estimation of these functions in randomly inhomogeneous media is a formidable task and has been addressed in the literature [43–45].

The formulation given above is applicable for immiscible fluids. If fluids mix, the governing equations become more complicated owing to mass fluxes of one fluid into another [46]. The mass balance equation is itself modified by factors such as hydrodynamic dispersion, requiring significant reformulation of the equations of mass, momentum, and energy.

Initial and boundary conditions: The system of equations given above is coupled and can only be numerically solved. Simulation studies with such systems have been discussed in the literature by various authors [47–51]. Specific solutions can be obtained in selected geometries with known boundary conditions. Two configurations, one and two dimensional, respectively, are sketched in Fig. 2.6. In the one-dimensional simulation, oil and water pressures are specified on each side of the physical domain, temperature is prescribed on the left side, and a suitable *outflow* condition is applied on the right. Initial conditions for the phase pressures and temperature over the entire domain are also required. In two-dimensional simulation, the figure sketched is a quarter of a repeating geometry of injection and production wells. The former carries pressurized and heated water, while oil and water are recovered at the other end. Thus, pressures and temperature are prescribed at the injection well, shown in Fig. 2.6 as a quarter of a circle. Pressures and a suitable temperature outflow condition are prescribed at the production well. On other boundaries, the symmetry condition in the form of a zero normal derivative boundary condition is prescribed.

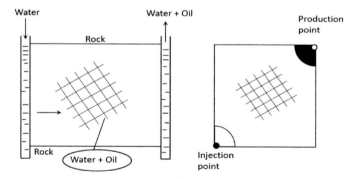

Fig. 2.6 One (left)- and two (right)-dimensional geometries chosen for simulation of enhanced oil recovery. In a two-dimensional geometry, water is injected at the injection well shown as a quarter-circle. Oil and water are recovered at the production well

2.6.2 Multiphase Multicomponent Flow

Generalizing oil–water flow in a porous medium, it is possible to imagine a scenario where gas, liquid, and solid fill the pore space while each phase has multiple chemical entities present. This is an example of multiphase, multicomponent flow and transport in a porous medium. As a specific example, a non-isothermal, multiphase, multispecies model is presented here for the simulation of CH_4 gas production via depressurization of a gas hydrate reservoir and simultaneous CO_2 injection [52–57]. The subject is discussed in detail in Chap. 7.

The porous medium in the present discussion is soil beneath water bodies such as lakes and oceans. In the pore space, all three phases—solid, liquid, and gas—are present. Each phase contains multiple species such as water and methane. The species balance equation, for example, of methane, is to be summed up over all the phases. The phase equilibrium relation brings in thermodynamic aspects in transport modeling. An example is the amount of methane present in the gas hydrate (solid) phase at a given pressure and temperature. Departure from equilibrium will result in reaction kinetics that will proceed at a speed in proportion to the distance from the equilibrium curve. The reaction chemistry is accompanied by stoichiometry, as discussed in Chap. 7 in the context of gas hydrates. In principle, each phase has its own pressure, temperature, and velocity. In the following discussion, the gas phase is treated to be ideal, local temperature to be equal among the phases including the solid matrix (thermal equilibrium) and solid phase velocity (soil as well as the hydrate deposits in the pores) to be zero.

The governing equations are written for simplicity in a one-dimensional Cartesian geometry bounded by the depressurization well at one end where methane is recovered and the injection well at the other where pressurized CO_2 can be introduced. The present model accounts for transport phenomena in the five components, namely aqueous, gas, CH_4-hydrate, CO_2-hydrate, and the geologic media, spread over three phases. While the first two components are mobile, the other three are treated as immobile. Such a distinction helps in accounting for the thermophysical properties of CH_4- and CO_2-hydrates. At a time instant and a given location, all five components present in the system are assumed to be in thermal equilibrium with each other. Temperature itself varies considerably over the spatial extent of the reservoir and is obtained as a part of the overall solution. The gas phase comprises a mixture of CH_4 and CO_2, while the liquid phase is aqueous and contains only H_2O. The solubility of CH_4 and CO_2 gases in the aqueous phase and the presence of H_2O in the gas phase can be shown to have small impact on gas recovery and is neglected.

Let subscripts 'g', 'mh', and 'ch' indicate gas, methane hydrate, and CO_2 hydrate, respectively. Superscripts 'm', 'c', and 'w' indicate methane, CO_2, and water, respectively. Let ω refer to mass fraction; thus ω_g^m stands for the mass fraction of methane in the gas phase. Similarly, the symbol ω_c^m is the mass fraction of CO_2 in the gas phase. Since these are the only two components of the gas phase, $\omega_g^m + \omega_c^m = 1$. Noting that methane is present in the gas phase as well as in methane hydrate (a solid phase), a mass balance equation of methane cutting across phases will read as follows:

$$\text{Methane} \quad \varepsilon \frac{\partial}{\partial t}\left(\rho_g s_g \omega_g^m + \rho_{mh} s_{mh} \omega_{mh}^m\right) = -\frac{\partial}{\partial x}\left(\rho_g \omega_g^m v_g + J_g^m\right) + \dot{m}_g^m + \dot{m}_{mh}^m$$

(2.52)

The left side is a mass storage term of methane in the REV. The first term on the right includes the diffusive mass flux given by symbol J (Eq. 2.54 below) and the advective mass flux that depends on the gas speed. Both terms serve to transport methane through the porous region. The velocity field is assumed to be Darcian, and the momentum equation of each fluid phase (gas and liquid) is expressed in one-dimensional form as:

$$v_\gamma = -\frac{K_{abs} k_{r\gamma}}{\mu_\gamma} \frac{\partial P_\gamma}{\partial x}, \gamma = g, l$$

(2.53)

The diffusive mass fluxes of methane and CO_2 appearing in the gas phase mass transport equations are given by the Fick's law, namely:

$$J_g^i = -\phi \rho_g s_g D_g^i \frac{M^i}{M_g} \nabla(\chi_g^i), \quad i = m, c$$

(2.54)

The source terms in the mass conservation equation of methane include possible chemical reactions in the gas phase that produce methane $\left(\dot{m}_g^m\right)$ and formation/dissociation reactions that generate/consume methane from the hydrates in the solid phase $\left(\dot{m}_{mh}^m\right)$.

Extending this approach to CO_2, the mass conservation equation is written as

$$\phi \frac{\partial}{\partial t}\left(\rho_g s_g \omega_g^c + \rho_{ch} s_{ch} \omega_{ch}^c\right) = -\frac{\partial}{\partial x}\left(\rho_g \omega_g^c v_g + J_g^c\right) + \dot{m}_g^c + \dot{m}_{ch}^c$$

(2.55)

The mass conservation equation of water that constitutes the liquid phase is similarly written as:

$$\phi \frac{\partial}{\partial t}\left(\rho_l s_l \omega_l^w + \rho_{mh} s_{mh} \omega_{mh}^w + \rho_{ch} s_{ch} \omega_{ch}^w\right) = -\frac{\partial}{\partial x}\left(\rho_l \omega_l^w v_l\right) + \dot{m}_l^w + \dot{m}_{mh}^w + \dot{m}_{ch}^w$$

(2.56)

The symbols carry their usual meaning. From left to right, the right side of Eqs. 2.55 and 2.56 carries the advection velocity, diffusive mass flux, and mass source terms arising from chemical reactions.

Using a volume-averaged temperature over the REC covering the three phases, the thermal energy equation is written as:

$$\frac{\partial}{\partial t}\left[\sum_{\gamma=l,g,mh,ch}\left(\phi\rho_\gamma s_\gamma U_\gamma\right)+(1-\phi)\rho_s U_s\right]+\sum_{\gamma=l,g}\left[\sum_{i=m,c,w}\frac{\partial}{\partial x}\left(v_\gamma\rho_\gamma s_\gamma \omega_\gamma^i H_\gamma^i\right)\right]$$

$$+\sum_{i=m,c}\frac{\partial}{\partial x}\left(J_g^i H_g^i\right)=\frac{\partial}{\partial x}\left(K_{eq}\frac{\partial T}{\partial x}\right)+\sum_{\gamma=l,g,mh,ch}\left[\sum_{i=m,c,w}\frac{\partial}{\partial x}\left(\dot{m}_\gamma^i H_\gamma^i\right)\right]+\dot{E}$$

$$(2.57)$$

The index γ sums up contributions from the three phases; the index 'i' performs summation over the applicable species. Symbol $\dot{E}$ refers to heat released/absorbed during exothermic and endothermic reactions in the porous region. Symbol U is velocity with the solid phase velocity U_s being, usually, zero. Symbol H is used to indicate phase enthalpy of the appropriate species.

The closure of the set of governing equations requires applicable constitutive relations and reaction kinetics. Specific forms of the constitutive relations, kinetic data, and heat release function are given in Chap. 7.

2.7   Mass Transfer

Saltwater mixing with freshwater in soil and leaching of chemicals from buried containers are examples of mass transfer in porous media. The transport phenomenon referred here is of a solute, for example, salt in a solvent such as water from regions of high concentration toward the low. It is possible to envision this transfer as analogous to thermal energy, temperature being equivalent to solutal concentration [2, 5, 7–9, 58–60]. In this discussion, concentration (C; also called, *mass* concentration) is expressed in units of mass of the solute per unit volume of the solution (e.g., mg/l). The solution in the present discussion is *clear*, namely free of particulates. In a clear medium, diffusive mass fluxes relate to the concentration gradient. For low solute concentrations (*dilute* solution), the relationship is linear with the proportionality constant being given by the mass diffusivity of the solute in the solution. This treatment is quite similar to the heat flux—temperature gradient relationship contained in Fourier's law of heat conduction. In the mass transfer context, the linear relation, namely Fick's law, is written as

$$\dot{m}'' = -D\nabla C \qquad (2.58)$$

Equation 2.58 connects mass flux (units of kg/m^2-s) and the concentration gradient via the mass diffusivity, D (units of m^2/s). Mass units can be equivalently expressed in molar form.

Consider mass transfer of a dissolved species in a flowing fluid. The physical law that governs the transport of the solute is the conservation of mass. In a clear medium, it leads to a mathematical form that is similar to thermal energy transport [59, 60] and is given as follows:

$$\left(\frac{\partial C}{\partial t} + \mathbf{u} \cdot \nabla C\right) = \nabla \cdot D\nabla C$$

In dilute solutions, diffusivity is practically independent of concentration, and the right-hand side can be expressed as $D\nabla^2 C$.

When the solutal mass balance equation is extended to a porous medium, mass diffusivity is replaced by the mass dispersion tensor D_m as in Eq. 2.32. Hence, for a porous medium

$$\left(\frac{\partial C}{\partial t} + \frac{\mathbf{u}}{\varepsilon} \cdot \nabla C\right) = \nabla \cdot [D_m]\nabla C \qquad (2.59)$$

Here, $\mathbf{u}$ is Darcian velocity. Equation 2.59 assumes that solute transfer is in the liquid phase, with the solid media playing merely a passive role of creating tortuosity in the flow path. In this respect, Eq. 2.59 is analogous to the thermal equilibrium model discussed in Sect. 2.5.

Non-equilibrium effects as in Sect. 2.5.1 can appear in mass transfer if the solute precipitates out of the solution and deposits over the solid surface or is *adsorbed* by it. As a result, the solute fraction in the solution progressively diminishes in the flow direction. In many instances, the interest is in the solute concentration entirely in the fluid phase. The redistribution of salt within the solid particles may not be as relevant. In any case, mass transfer rates of salts in a solid phase are exceptionally slow, and the salt may be taken to reside entirely at the surface. The role of the solid phase in mass transfer is then to introduce purely a source term in Eq. 2.59. If the solid phase is assumed to remain clean at all times, the source term in the mass balance equation will be proportional to the local solute concentration. Hence, for a solute getting adsorbed into the solid phase

$$\left(\frac{\partial C}{\partial t} + \frac{\mathbf{u}}{\varepsilon} \cdot \nabla C\right) = \nabla \cdot [D_m]\nabla C - \beta C \qquad (2.60)$$

The adsorption and dispersion parameters β and D_m have to be estimated independently from experiments. The adsorption parameter β may be estimated from reaction chemistry under certain circumstances.

If a model for the transport of solid phase concentration C^s is available, Eq. 2.60 can be rewritten for the solute concentration in the fluid phase $\left(C^f\right)$ as

$$\left(\frac{\partial C^f}{\partial t} + \frac{\mathbf{u}}{\varepsilon} \cdot \nabla C^f\right) = \nabla \cdot [D_m]\nabla C^f - \beta(C^f - C^s) \qquad (2.61)$$

Equation 2.61 includes the following special cases:

(a) mass transfer of the solute from the solution to a clean matrix ($C^s = 0$ everywhere);

(b) mass transfer from a contaminated matrix to a clean solution ($C^s = 0$ on the inflow plane); apart from

(c) the general adsorption–desorption problem of contaminant transport in a porous medium.

Initial and boundary conditions for Eqs. 2.60 and 2.61 will depend on the physical problem being studied. For example, a salt solution entering a water-saturated porous medium will be subject to the following conditions:

$$
\begin{aligned}
t = 0 \quad & C^f(x,t) = 0 \, C^s(x,t) = 0 \quad x > 0 \\
x = 0 \quad & C^f(x,t) = C_0^f C^s(x,t) = 0 \, t > 0 \\
x \to \infty \quad & C^f(x,t) = 0 \, C^s(x,t) = 0 \, t > 0
\end{aligned}
\tag{2.62}
$$

Revised forms of Eq. 2.61 can be further developed for the following situations:

i. The transported species is radioactive with a decay constant d_λ. Here, the adsorption parameter β in Eq. 2.60 is replaced by $\beta + d_\lambda$.

ii. Species concentration is large, the dilute solution approximation is not valid, and diffusivity is a function of concentration.

iii. Dispersion models analogous to Eq. 2.32 are introduced.

iv. The incoming solution and the porous medium are at distinct temperatures, leading to a combined heat and mass transfer problem.

v. *Gas phase*: When gas (such as methane) is adsorbed on to the surface of the particles (of coal), the extent of adsorption depends on the local gas pressure. Thus, the adsorption–desorption process is determined by the difference in pressures between the gas phase and in the pores. For a gaseous medium, concentration can be related to pressure using the ideal gas law. With a and b as material-specific parameters, the concentration–pressure relation is given by the Langmuir equation

$$
C = \frac{ap}{p+b}
$$

2.8   Combined Heat and Mass Transfer

In a class of applications related to drying of an initially wet porous region, temperature gradients are jointly present with gradients of moisture content. Moisture here is present in the pore space in liquid form. With the bulk velocity taken to be zero, the mass flow rate and momentum equations need not be separately considered. The porous medium is taken to be homogeneous and isotropic for the present discussion, gravity effects are neglected, and the local thermal equilibrium model is employed. Humidity is analogous to solute concentration considered in Sect. 2.7. The respective heat and mass fluxes are modified forms of the Fourier and Fick's laws of diffusion for heat and moisture migration and are given as

$$\dot{q}'' = -k_m \nabla T$$
$$\dot{m}'' = -D_m \nabla H - D_T \nabla T \tag{2.63}$$

The model parameters of Eq. 2.63 are positive and are to be determined from experiments. Symbol D_T stands for the diffusion coefficient of mass transfer arising from a temperature gradient while k_m and D_m are the usual heat and mass transfer conductivity/diffusivity, respectively. For small evaporation rates from the liquid phase, Chang and Weng [61] have shown that these can be determined from principles of thermodynamics; also see [62].

The heat and mass transfer equations now solve for temperature (T) and moisture content (H) in the pore space and are given as:

$$(\rho C)_m \frac{\partial T}{\partial t} = k_m \nabla^2 T + Q$$
$$\frac{\partial H}{\partial t} = D_m \nabla^2 H + M \tag{2.64}$$

Here, suffix 'm' refers to the effective medium properties of the porous region. Source terms Q and M refer to heat transfer arising from moisture migration primarily related to latent heat release and mass transfer related to possible condensation as well as the superimposed temperature gradient.

In the physical model, the solid phase, for example, wet sand dries during flow of dry air over it, changing its moisture and energy content. Thus, the energy source term can be written as

$$Q = (\rho h_{lv})_m \frac{\partial H}{\partial t}$$

Here, h_{lv} is the latent heat of evaporation of moisture at the local medium temperature. The mass source term in the moisture content equation is the divergence of mass flux associated with the temperature gradient and is expressed in one-dimensional form as

$$M = \frac{\partial}{\partial x}\left(D_T \frac{\partial T}{\partial x}\right)$$

Equation 2.64 is a coupled system of partial differential equations in terms of moisture content and temperature and can be numerically solved, subjected to suitable initial and boundary conditions.

Consider an initially wet, cold, horizontal, one-dimensional porous bed that is subjected to a temperature difference. Hence, the initial condition is:

$$t = 0 \quad T(y, t) = T_C; \quad H(y, t) = 1$$

The thermal boundary conditions are:

$$T(y = 0, t) = T_C; \quad T(y = L, t) = T_H$$

In many contexts, the lower surface is cooler to prevent natural convection effects.

The top surface is exposed to dry air from where water evaporates at a rate depending on the local mass transfer coefficient (h_m). Hence

$$-D_m \frac{\partial H}{\partial y}\bigg|_{y=L} = h_m(H(L, t) - 1)$$

Assuming the lower boundary to be impermeable, the mass flux here is zero, leading to:

$$-D_m \frac{\partial H}{\partial y}\bigg|_{y=0} - D_T \frac{\partial T}{\partial y}\bigg|_{y=0} = 0$$

For large evaporation rates, the local pressure field is disturbed and a Darcian velocity component is to be added to the mass transfer equation. For a discussion on the drying of a porous medium using high temperature steam, see [63].

The above formulation can be extended to a porous medium filled with moist air and subjected to a cold boundary condition on one side. Under suitable conditions, water will condense on the cold side where it is drained, and the sign of vapor transport is reversed from the hot side toward the cold.

2.9 Flow, Heat, and Mass Transfer

In several applications, flow may arise from density differences in the fluid region, density being a function of temperature and species concentration. An example of such an application, shown in Fig. 2.7, is freshwater overlaid on saltwater, the overall arrangement subjected to a temperature difference [64–66]. Terms such as double-diffusive and triple-diffusive convection are often used in this context. For the present discussion, a thermal equilibrium model is adopted for heat transfer. The solid phase of the porous region is taken to be non-participating in mass transfer, solutal concentration referring to the fluid phase alone. The variation of fluid density with temperature and concentration can be assumed linear (within limits, [67]) and expressed as

Fig. 2.7 An example where
fluid flow is coupled with heat
and mass transfer in a porous
region. The walls are taken to
be impermeable to mass
transfer (mass flux = 0) while
temperatures are prescribed

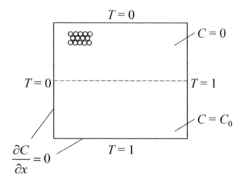

$$\rho = \rho_0(1 - \beta(T - T_0) + \gamma(C - C_0))$$

Here, suffix '0' indicates a reference state, for example, the initial undisturbed
condition of the fluid region. Symbols β and γ refer to the volumetric expansion
coefficient of the fluid relative to temperature (positive, K^{-1}) and concentration
(positive, $(mg/l)^{-1}$), respectively. It is expected that an increase in temperature will
lower density while increase in concentration will increase density further. Hence,
coefficients β and γ are expected to be positive quantities. Body force contributions
arise from distortions in temperature and concentration that perturb the density
field. Changes in density and other fluid properties are neglected in other terms, for
example, inertia and viscous. Such an approach is called the *Boussinesq* approximation [67].

With z as a coordinate oriented in the vertical direction, we define a modified
pressure for ease of mathematical formulation as follows:

$$\tilde{p} = p + \rho_0 g z$$

The mass, momentum, energy, and solutal transport equations can be jointly
written as:

$$\nabla \cdot \mathbf{u} = 0$$

$$\frac{\rho_0}{\varepsilon}\left(\frac{\partial \mathbf{u}}{\partial t} + \frac{\mathbf{u}}{\varepsilon} \cdot \nabla \mathbf{u}\right) = -\nabla\tilde{p} + \rho_0 g[\beta(T - T_0) - \gamma(C - C_0)]\mathbf{k} - \frac{\mu}{K}\mathbf{u}$$

$$- \frac{\rho F}{\sqrt{K}}\mathbf{u}^2 + \frac{\mu}{\varepsilon}\nabla^2\mathbf{u}$$

$$(\rho C)_m \frac{\partial T}{\partial t} + (\rho C)_f \mathbf{u} \cdot \nabla T = \nabla \cdot [k_m]\nabla T$$

$$\left(\frac{\partial C}{\partial t} + \frac{\mathbf{u}}{\varepsilon} \cdot \nabla C\right) = \nabla \cdot [D_m]\nabla C$$

(2.65)

In Eq. 2.65, **k** is a unit vector in the vertically upward direction. Note that temperature and concentration effects are opposed to each other in the sense that concentration variation as in Fig. 2.7 can diminish velocities and stabilize the flow field.

For the problem shown in Fig. 2.7, the initial and boundary conditions are:

$$t = 0 \, \mathbf{u} = 0, \, T = 0, \, C = F(z)$$
$$\text{walls}, \mathbf{u} = 0, T = \text{given}, \frac{\partial C}{\partial n} = 0$$

Here, n is the outward drawn normal on a solid surface. In Eq. 2.65, flow is set up due to inhomogeneities in the density field arising from a distribution of solutal concentration and temperature. In this respect, the system of equations in (2.65) is coupled and needs to be jointly solved.

2.10   Nanoscale Porous Media

Hydrocarbons (liquids as well as gases), water, and aqueous solutions can occupy pores and fractures of natural reservoirs whose length scales extend from a few millimeters down to nanometers. Pores in membranes used for distillation and component separation are in the nanometer range. The overall characteristics of the flow and transport processes depend on the connectivity of the pores over the length scales, and the details of the physical processes, as one proceeds from the macro- to the micropores in the porous medium. If there were to be no change in the physical laws underlying transport of mass, momentum, and energy, it would be appropriate to define effective properties for quantities such as permeability, conductivity, and dispersion and treat the porous region as a continuum. This assumption has been implicitly utilized in analysis when length scales varied over two or three orders of magnitude. The issue of whether it holds over a spread of 6–7 orders of magnitudes is as yet unresolved. Preliminary research shows that significant differences are to be anticipated. For example, small pore dimensions in the micron range will yield effective permeability greater than what is predicted by Eq. 1.3.

The flow of fluids, fluid–matrix interactions, chemical reactions, and wetting characteristics in a nano-environment are topics of research that have just begun to be reported in the literature [68–72]. It calls for molecular level experiments as well as simulation to relate microscopic concepts of surface texture, three-phase contact line, and probability distribution functions to the macroscopic, such as temperature and pressure. The determination of effective medium properties in the presence of nanoscale phenomena is the next challenge in the study of transport phenomena in porous media.

2.11   Multiscale Porous Media

The heterogeneity of a porous medium can be accounted for in the mathematical model by treating quantities such as permeability and dispersion coefficient as space-dependent model parameters that are determined from carefully calibrated inverse experiments. Meaningful predictions are possible when permeability variations are continuous and within limits. This approach is, however, unsatisfactory from a theoretical viewpoint, when the degree of heterogeneity is quite high. This is because the model parameters thus determined are likely to be specific to the application considered; in addition, they could be scale- and phenomenon-dependent as well.

A greater generality can be built into the theory of transport phenomena in porous media, by assigning a mathematical structure to the heterogeneity of the physical region. The fractal structure is an example, wherein the pore scales have a continuous variation in the wave number space and equivalently show an infinite correlation length in the physical space [73, 74]. The macroscopic properties now depend on the fractal dimensions, along with adjustable coefficients that are to be determined from experiments.

The pore geometry of heterogeneous porous media may not have a fractal structure but can show a hierarchical variation in many applications. As described by Cushman and coworkers [75–77], this indicates that the macroscopic medium can be decomposed into successively nested, interacting physical subunits. The decomposition can be discrete or continuous; the available literature has examined a finite number of hierarchies in the porous region. It is possible for the physical phenomena at every ladder of the hierarchy to be identical, except that they develop inter-dependence at their interfaces. Alternatively, the physical dimensions may be so far apart that transport processes in the hierarchy are quite distinct. For example, it is conceivable that the smallest length scales fall in the micro- or the nanorange, while the bulk of the porous region is at the macroscale. This is an example of functional hierarchy in transport processes coexisting with the structural hierarchy of the porous region. For a discussion on transport in a two-level hierarchy, dual porosity system, see [78].

The differences in the volume-averaged governing equations between homogeneous and hierarchical porous media become clear from the following discussion. A scalar function is denoted as $F(x)$ in the space variable x while $\mathbf{F}(x)$ is a vector-valued function. For a homogeneous region, Gauss divergence theorem can be written for the scalar- and vector-valued functions, respectively, as follows:

$$\int_\omega \nabla F(x)dV = \int_{\partial\omega} F(x)n dS$$

$$\int_\omega \nabla \cdot \mathbf{F}(x)dV = \int_{\partial\omega} \mathbf{F}(x) \cdot n dS$$

$$(2.66)$$

Here, ω is a representative volume, and $\partial\omega$ is its surface, with n being a unit outward drawn normal on the surface. When the above relations are applied to a heterogeneous medium, volume ω is replaced by ω_1, the volume occupied by all particles and pores at scale 1 (the corresponding hierarchy). In addition, $\partial\omega_1$ is the partial surface area at this scale, and $\partial\omega$ is the total surface area. Gauss's theorem applied to a single scale of the hierarchy can be derived as [79]:

$$\int_{\omega} \nabla F(x)dV = \nabla \int_{\omega_1} F(x)dV + \int_{\partial\omega - \partial\omega_1} F(x)ndS$$

$$\int_{\omega} \nabla \cdot \mathbf{F}(x)dV = \nabla \cdot \int_{\omega_1} \mathbf{F}(x)dV + \int_{\partial\omega - \partial\omega_1} \mathbf{F}(x) \cdot ndS$$

(2.67)

The additional terms in the above equations (relative to Eq. 2.66) significantly increase the mathematical complexity of the governing equations.

The process of integrating the governing equations over a subdomain of the physical space can give rise to constitutive variables that have a *non-local* character. It will require parameter estimation experiments to be conducted at appropriate scales, and not the macroscale alone. These are additional complications in the study of flow and transport in hierarchical porous media.

Vankan et al. [80] derived equations for blood perfusion in a biological tissue using the hierarchical mixture theory. While blood is the fluid of interest, the pores in the tissue form on the lowest scale of the porous medium. At a larger length scale, a few larger arteries supply blood through a diverging, arteriolar vascular channel. Blood reaches the capillaries and is subsequently drained by the converging veins. Blood vessels are numerous, uniformly distributed, and occupy a large part of the hierarchy. Geometric and mechanical properties of the vascular structure such as diameter, wall thickness and stiffness, orientation and density depend on the hierarchy being addressed. Consequently, the flow field is a function of position on one hand and exhibits on the other, a dependence on the scale at which the flow takes place.

The mathematical formulation for transport of heat and moisture in an unsaturated porous medium with three spatial scales is developed in [77]. It is shown here that the number of variables that need to be specified via constitutive relationships is over 30! Thus, a general formulation of transport in continuous hierarchical porous media has considerable challenge.

2.12 Closure

Derivation of equations governing flow, heat, and mass transfer in porous media is discussed in the present chapter. The starting point is Darcy's law that can be gradually extended to higher flow rates, complex pore geometries, and the appearance of interfaces of immiscible fluids. A second starting point is the system of equations valid for a homogeneous fluid medium; it can be generalized to a multiphase system such as a porous medium by introducing source terms and effective medium properties. In each approach, the model carries a large number of parameters that are sensitive to the pore structure, though to a lesser extent on the thermophysical properties of the constituent media. Thus, success in modeling transport in porous media is linked to careful parameter estimation from experiments. This step is expected to become critical in multiscale porous media where the pore scales span several orders of magnitude. Theoretical derivation of the applicable form of the governing equations may prove to be advantageous under these circumstances. Modeling flow in the non-continuum regime is discussed in Chap. 3.

References

1. J. Bear, *Dynamics of Fluids in Porous Media* (American Elsevier Pub. Co., New York, 1972)
2. J. Bear, Y. Bachmat, *Introduction to Modeling of Transport Phenomena in Porous Media* (Kluwer, Dordrecht, 1990)
3. S. Whitaker, *The Method of*, vol. Averaging (Kluwer Academic Publishers, Boston, 1999)
4. G.F. Pinder, *Essentials of Multiphase Flow and Transport in Porous Media* (Wiley, Hoboken, 2008)
5. M. Kaviany, *Principles of Heat Transfer in Porous Media* (Mechanical Engineering Series, Springer, New York, 1991)
6. K. Vafai (ed.), *Handbook of Porous Media*, 3rd edn. (CRC Press, Boca Raton, USA, 2015)
7. J. Bear, A. Verruijt, *Modeling Groundwater Flow and Pollution* (Reidel, Dordrecht, 1987)
8. D.B. Ingham, I. Pop, *Transport Phenomena in Porous Media* (Pergamon Press, Oxford, 2002)
9. D.A. Nield, A. Bejan, *Convection in Porous Media*, 4th edn. (Springer, New York, 2013)
10. M. Le Ravaled-dupin, *Inverse Stochastic Modeling of Flow in Porous Media* (Technip, Paris, 2005)
11. F. White, *Fluid Mechanics*, 2nd edn. (McGraw-Hill, New York, 1986)
12. F.W. Schwartz, H. Zhang, *Fundamentals of Groundwater* (Wiley, New York, 2003)
13. C.R. Fitts, *Groundwater Science* (Academic Press, London, 2002)
14. D.W. Hahn, M.N. Ozisik, *Heat Conduction*, 3rd edn. (Wiley, New Jersey, 2012)
15. M. Sahimi, *Flow and Transport in Porous Media and Fractured Rock* (Wiley, New Jersey, 2011)
16. A. Amiri, K. Vafai, Analysis of dispersion effects and non-thermal equilibrium, non-Darcian, variable porosity incompressible flow through porous media. Int. J. Heat. Mass Transf. **37**, 939–954 (1994)
17. M. Sozen, T.M. Kuzay, Enhanced heat transfer in round tubes with porous inserts. Int. J. Heat Fluid Flow **17**, 124–129 (1996)
18. A.V. Kuznetsov, K. Vafai, Analytical comparison and criteria for heat and mass transfer models in metal hydride packed beds. Int. J. Heat Mass Transf. **38**, 2873–2884 (1995)
19. C.A.J. Fletcher, Computational Techniques for Fluid Dynamics, Vol. I and II, Springer, New York, 1988
20. J.D. Anderson Jr., *Modern Compressible Flow with Historical Perspective* (McGraw-Hill, New York, 1990)

21. P.G. Drazin, W.H. Reid, *Hydrodynamic Stability* (Cambridge University Press, Cambridge, UK, 1981)
22. M.J.S. de Lemos, M.H.J. Pedras, Recent mathematical models for turbulent flow in saturated rigid porous media. ASME Fluids Engg. **123**(4), 935–940 (2001)
23. Y. Takatsu, T. Masuoka, Turbulent phenomena in flow through porous media. J. Porous Media **1**, 243–251 (1998)
24. M.J.S. de Lemos, *Turbulent Impinging Jets into Porous Materials* (Springer Briefs in Computational Mechanics, New York, 2012)
25. P. Holmes, J.J. Lumley, G. Berkooz, C.W. Rowley, *Turbulence, Coherent Structures, Dynamical Systems, and Symmetry* (Cambridge University Press, Cambridge, 2012)
26. H. Schlichting, Boundary-layer Theory, McGraw-Hill, New York, 1979 (8th edition: 2000)
27. D. Getachew, W.J. Minkowycz, and J.L. Lage, A modified form of the k-ε model for turbulent flow of an incompressible fluid in porous media. Int. J. Heat Mass Transfer, pp. 2909–2915, 2000
28. W.M. Kays, M.E. Crawford, *Convective Heat and Mass Transfer*, 3rd edn. (McGraw-Hill, New York, 1993)
29. S.M. Ghiaasiaan, *Convective Heat and Mass Transfer* (Cambridge University Press, Cambridge, 2011)
30. L.C. Davis, B.E. Artz, Thermal conductivity of metal-matrix composites. J. Appl. Phys. **77** (10), 4954–4958 (1995)
31. G. Degan, *Flow and Transport in Porous Formations* (Springer-Verlag, Berlin, 1989)
32. C.T. Hsu, P. Cheng, Thermal dispersion in a porous medium. Int. J. Heat Mass Transf. **33**, 1587–1598 (1990)
33. A. Levy, S. Sorek, G. Ben-Dor, J. Bear, Evolution of the Balance Equations in Saturated Thermoelastic Porous Media following abrupt simultaneous changes in Pressure and Temperature. Transp. Porous Media **21**, 241–268 (1998)
34. A. Levy, G. Ben-Dor, S. Sorek, Numerical Investigation of the Propagation of Shock waves in Rigid Porous Materials: Flow Field Behaviour and Parametric Study. Shock Waves **8**, 127–137 (1998)
35. A. Hunt, R. Ewing, *Percolation Theory for Flow in Porous Media* (Springer, Berlin, 2009)
36. C.M. Case, *Physical Principles of Flow in Unsaturated Porous Media* (Oxford University Press, Oxford, 1994)
37. MTh van Genuchten, A Closed-Form Equation for Predicting the Hydraulic Conductivity of Unsaturated Soils. Soil Sci. Soc. Am. J. **44**(5), 892–898 (1980)
38. A. Szymkiewicz, *Modelling Water Flow in Unsaturated Porous Media* (Springer, New York, 2013)
39. T.C. Boberg, *Thermal Methods of Oil Recovery: An Exxon Monograph* (Wiley, New York, 1988)
40. R.E. Ewing, *The Mathematics of Reservoir Simulation* (SIAM, Philadelphia, 1983)
41. P. Daripa, J. Glimm, B. Lindquist, M. Maesumi, O. McBryan, On the simulation of heterogeneous petroleum reservoirs, in *Numerical Simulation in Oil recovery*, vol. 11, ed. by M.F. Wheeler, The IMA Volume in Mathematics and its Applications, (Springer-Verlag, 1986), pp 89–103
42. K. Aziz, Modeling of thermal oil recovery processes, in *Mathematical and Computational Methods in Seismic Exploration and Reservoir Modeling*, ed. by W.E. Fitzgibbon (Philadelphia, SIAM, 1986), pp. 3–17
43. D.S. Oliver, Multiple realizations of the permeability field from well-test data. Soc. Petrol. Eng. J. **1**(2), 145–154 (1996)
44. A.C. Reynolds, N. He, L. Chu, D.S. Oliver, Reparameterization techniques for generating reservoir descriptions conditioned to variograms and well-test pressure data. Soc. Petrol. Eng. J. **1**(4), 413–426 (1996)
45. F. Zhang, A.C. Reynolds, D.S. Oliver, Evaluation of the reduction in uncertainty obtained by conditioning a 3D stochastic channel to multiwell pressure data. Math. Geol. **34**(6), 715–742 (2002)

46. F. Civan, Implications of alternate macroscopic descriptions illustrated by general balance and continuity equations. J. Porous Media **5**(4), 271–282 (2002)
47. P.K.W. Vinsome, Fully implicit versus dynamic implicit reservoir simulation. J. Canadian Petro. Tech., 49–82, 1985
48. R. Lenormand, E. Touboul, C. Zarcone, Numerical models and experiments on immiscible displacements in porous media. J. Fluid Mech. **189**, 165–187 (1988)
49. C.T. Tan, G.M. Homsy, Simulation of nonlinear viscous fingering in miscible displacement. Phys. Fluids **31**(6), 1330–1338 (1988)
50. M.D. Stevenson, M. Kagan, W.V. Pinczewski, Computational methods in petroleum reservoir simulation. Comput. Fluids **19**, 1–19 (1991)
51. H.R. Zhang, K.S. Sorbie, N.B. Tsibuklis, Viscous fingering in five-spot experimental porous media: new experimental results and numerical simulation. Chem. Eng. Sci. **52**, 37–54 (1997)
52. G. Ahmadi, C. Ji, D.H. Smith, Numerical solution for natural gas production from methane hydrate dissociation. J. Petrol. Sci. Eng. **41**, 269–285 (2004)
53. G.C. Fitzgerald, M.J. Castaldi, Y. Zhou, Large scale reactor details and results for the formation and decomposition of methane hydrates via thermal stimulation dissociation. J. Petrol. Sci. Eng. **94**, 19–27 (2012)
54. G.J. Moridis, T.S. Collett, R. Boswell, M. Kurihara, M.T. Reagan, C.A. Koh, Toward production from gas hydrates: current status, assesment of resources, and simulation-based evaluation of technology and potential. SPE Reserv Eval. Eng. **12**(5), 745–771 (2009)
55. E.D. Sloan Jr., C.A. Koh, Clathrate Hydrates of Natural Gases 3rd Edition, Series: Chemical Industries, Vol. 119, CRC Press, 2007
56. M. Uddin, D.A. Coombe, D. Law, W.D. Gunter, Numerical studies of gas hydrate formation and decomposition in a geological reservoir. ASME J Energy Resour Technol. **130**(3), 10–17 (2008)
57. M.D. White, S.K. Wurstner, B.P. McGrail, Numerical studies of methane production from class 1 gas hydrate accumulations enhanced with carbon dioxide injection. Mar. Pet. Geol. **28**(2), 546–560 (2011)
58. F. Civan, *Porous Media Transport Phenomena* (Wiley, New Jersey, 2011)
59. D. Ingham, A. Bejan, E. Mamut, I. Pop (eds.), *Emerging Technologies and Techniques in Porous Media* (Springer, New York, 2012)
60. J.M.P.Q. Delgado (ed.), *Heat and Mass Transfer in Porous Media* (Springer, New York, 2012)
61. W.J. Chang, C.I. Weng, An Analytical Solution to Coupled Heat and Moisture Diffusion Transfer in Porous Materials. Int. J. Heat Mass Transf. **43**, 3621–3632 (2000)
62. K.K. Khankari, R.V. Morey, S.V. Patankar, Mathematical Model for Moisture Diffusion in Stored Grain due to Temperature Gradients, Transactions of ASME –. J. Heat Transf. **37**(5), 1591–1604 (1994)
63. P. Perre, M. Moser, M. Martin, Advances in Transport Phenomena during Convective Drying with Superheated Steam and Moist Air. Int. J. Heat Mass Transfer **36**(1), 2725–2746 (1993)
64. A. Mojtabi, M.C. Charrier-Mojtabi, Double diffusive convection in porous media, in *Handbook of Porous Media*, ed. by K. Vafai (Taylor and Francis, New York, 2005), pp. 269–320
65. A.V. Kuznetsov, D.A. Nield, The Cheng-Minkowycz problem for the double-diffusive natural convective boundary layer flow in a porous medium saturated by a nanofluid. Int. J. Heat Mass Transf. **54**, 374–378 (2011)
66. S. Rionero, Triple diffusive convection in porous media. Acta Mech. **224**, 447–458 (2013)
67. B. Gebhart, Y. Jaluria, R.L. Mahajan, B. Sammakia, *Buoyancy-Induced Flows and Transport* (Springer, New York, 1988)
68. S. Roy, R. Raju, H.F. Chuang, B.A. Crudent, M. Meyyappan, Modeling gas flow through microchannels and nanopores. J. Appl. Phys. **93**(8), 4870–4879 (2003)
69. V.P. Sokhan, D. Nicholson, N. Quirke, Fluid flow in nanopores: an examination of hydrodynamic boundary conditions. J. Chem. Phys. **115**(8), 3878–3887 (2001)

70. M. Gad-el-Hak, The fluid mechanics of microdevices—The Freeman Scholar lecture. ASME J. Fluids Eng. **121**, 6–33 (1999)
71. G.A. Bird, *Molecular Gas Dynamics and the Direct Simulation of Gas Flows* (Clarendon Press, Oxford, 1994)
72. G. Karniadakis, A. Beskok, *Micro Flows-Fundamentals and Simulation* (Springer, New York, 2002)
73. R. Pitchumani, B. Ramakrishnan, A fractal geometry model for evaluating permeabilities of porous performs used in liquid composite molding. Int. J. Heat Mass Trans. **42**, 2219–2232 (1999)
74. S.W. Wheatcraft, G.A. Sharp, S.W. Tyler, Fluid flow and transport in fractal heterogeneous porous media, Chapter XI in *Dynamics of Fluids in Hierarchical Porous Media*, ed. By J.H. Cushman, Academic Press, New York, 1990
75. J.H. Cushman (ed.), *Dynamics of Fluids in Hierarchical Porous Media* (Academic Press, New York, 1990)
76. J.H. Cushman, *The Physics of Fluids in Hierarchical Porous Media: Angstroms to Miles* (Kluwer Academic, Dordrecht-Boston, 1997)
77. L.S. Bennethum, J.H. Cushman, Coupled solvent and heat transport of a mixture of swelling porous particles and fluids: single time-scale problem. Transp. Porous Media **36**, 211–244 (1999)
78. C.Y. Wang, C. Beckermann, Single vs. dual-scale volume averaging for heterogeneous multiphase systems. Int. J. Multiphase Flow **19**(2), 397–407 (1993)
79. V.S. Travkin, I. Catton, Transport phenomena in heterogeneous media based on volume averaging theory. Adv. Heat Transf. **34**, 1–144 (2001)
80. W.J. Vankan, J.M. Huyghe, M.R. Drost, J.D. Janssen, A. Huson, A finite element mixture model for hierarchical porous media. Int. J. Numer. Meth. Engg. **40**, 193–210 (1997)

Chapter 3
Mesoscale Interactions of Transport Phenomena in Polymer Electrolyte Fuel Cells

Partha P. Mukherjee, Ankit Verma, and Aashutosh Mistry

3.1 Introduction

Recent years have witnessed a tremendous interest in exploring fuel cells as a clean energy conversion system, owing to their high energy efficiency, environmental friendliness, and minimal noise [1–3]. Several fuel cell categories have evolved in the past decades, for example, direct methanol fuel cell (DMFC) [4], solid oxide fuel cell (SOFC) [5] with the polymer electrolyte fuel cell (PEFC) [6] emerging as the front-runner for automotive, portable, and stationary applications because of its low operating temperature, high-rated power density, fast start-up, high sensitivity to load variations, and relatively lower cost compared to the alternative versions of fuel cells.

A conventional PEFC, schematically shown in Fig. 3.1, utilizes the electrochemical conversion of hydrogen and oxygen into water to produce energy. The stratified structure consists of a proton (H^+)-conducting membrane sandwiched between two electrodes, anode and cathode, with individual current collectors providing the electrical connection to the load. Each of the electrodes comprises a flow channel directing the gas flow onto the porous gas diffusion layer (GDL) and the porous catalyst layer (CL), in that order. Hydrogen and air (humidified) stream form the feeds of the anode and cathode, respectively. The GDL serves two major purposes, facilitating the transport of the reactants and products, in and out of the electrochemically active CL and providing mechanical rigidity to the otherwise fragile electrode. Porous carbon structure of varying thicknesses is utilized for the GDL. CL comprises porous composites of polymer electrolyte binder (ionomer) and catalyst nanoparticles supported on a carbon substrate. Nano-sized platinum (Pt) or Pt alloys are used to catalyze the electrochemical reaction owing to their greater catalytic activity, higher surface area to weight ratio, and insensitivity to the acidic nature of the polymer membrane. Since electrochemical reactions occur only inside the CL, the carbon particles inside CL are also smaller as compared to those in GDL. Active area considerations rank higher in the CL, while improving the

© Springer International Publishing AG 2018

M.K. Das et al., *Modeling Transport Phenomena in Porous Media with Applications*, Mechanical Engineering Series, https://doi.org/ 10.1007/978-3-319-69866-3_3

transport properties such as porosity, tortuosity, and permeability is the major concern for GDL.

Hydrogen oxidation reaction (HOR) occurring at the anode CL exhibits orders of magnitude faster reaction kinetics as compared to the oxygen reduction reaction (ORR) occurring at the cathode CL [7]. Consequently, ORR is the primary source of voltage loss, and improving the cathode CL has been the focus of intense research efforts [8, 9]. Another reason for the cathode being a major limiting factor is the issue of efficient transport of water, the product of the ORR [10]. Optimal water content is required inside the membrane electrode assembly, deviation from which results in performance limitations. Low water content inhibits the proton conductivity of the membrane while excess water can block the porous pathways in the CL/GDL, causing hindered oxygen transport to the reaction sites, as well as

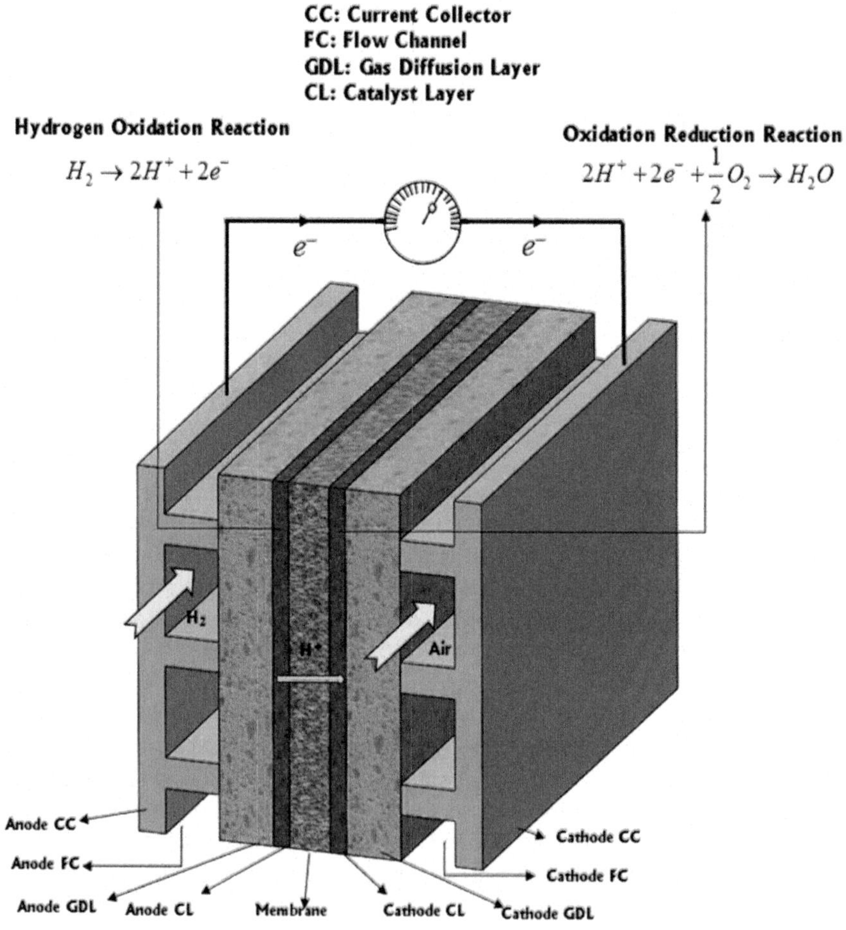

CC: Current Collector
FC: Flow Channel
GDL: Gas Diffusion Layer
CL: Catalyst Layer

Hydrogen Oxidation Reaction

$$H_2 \rightarrow 2H^+ + 2e^-$$

Oxidation Reduction Reaction

$$2H^+ + 2e^- + \frac{1}{2}O_2 \rightarrow H_2O$$

Fig. 3.1 Schematic diagram of a polymer electrolyte fuel cell

cover the electrochemically active sites in the CL rendering them inactive, and subsequently resulting in high surface overpotential. Therefore, improving the performance of PEFC necessitates a solution to the aforementioned issues.

The performance of a PEFC is characterized by the polarization curve, shown in Fig. 3.2, giving the relation between cell voltage and current density. The polarization curve shows three distinct voltage loss regimes, characterized by the dominant voltage loss mechanism, which is dependent on the magnitude of the cell current density. Low current density operation exhibits miniscule voltage loss, primarily due to the sluggish ORR at the cathode CL and is known as the *kinetic overpotential* regime. Transport limitations have not yet crept inside the system owing to low transport rates at small current densities. Intermediate current density operation is termed as the *ohmic overpotential* regime characterized by higher ionic transport resistance in the polymer electrolyte membrane and catalyst layer which dwarfs the kinetic limitations. Finally, high current density operation is dominated by mass transport limitations owing to the buildup of excessive liquid water which inhibits the efficient transport of reactant species (O_2) to the active sites in both GDL and CL as well as active surface passivation due to liquid water coverage. Thus, it is aptly termed as the *concentration overpotential* regime. The phenomenon of rapid elevation in water content at high current density is also known as 'flooding' which leads to the limiting current behavior in the fuel cell performance and cell shutdown.

Modeling of two-phase transport in PEFC electrode has been conventionally approached using two different methodologies, the macroscopic electrode scale

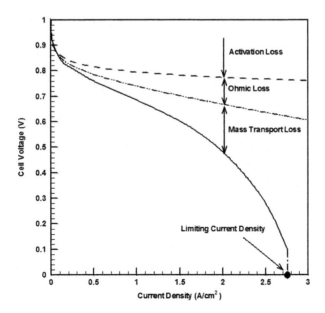

Fig. 3.2 Typical (performance) polarization curve of a PEFC [15]

models [11–14] using volume-averaged geometric and transport properties in porous electrode theory and the mesoscopic pore-scale models [15] involving solution of transport on fully resolved porous electrode microstructure incorporating all inhomogeneity. Macroscale models are advantaged by tractable computational cost but are limited by the use of empirical correlations and correction factors which cannot be completely discerned using experimental techniques. The intricate details of all the events happening at the pore scale are lost which necessitates the formulation of a mesoscale approach to unravel the structure–transport–performance interactions at the pore scale.

In this chapter, the analysis of mesoscale interactions occurring due to the underlying porous structure and transport phenomena in the catalyst layer and gas diffusion layer in a typical PEFC, is presented. In particular, the mesoscale modeling methodologies for two-phase transport inside PEFC, namely, lattice Boltzmann modeling (LBM) [16] and electrochemistry-coupled direct numerical simulation (DNS) [17] are highlighted in detail. The rest of the chapter is arranged as follows. First, a brief description of charge transport in porous media is presented. It follows by a summary of various models available for simulating pore-scale two-phase transport. The suitability of these models for application in PEFC is also discussed. Subsequently, a brief outline of microstructure reconstruction techniques is given which forms the precursor step to LBM and DNS studies. Finally, the LBM and DNS models are introduced. An exhaustive discussion of the model formulation and the results obtained in context of PEFCs are also presented.

3.2 Description of Charge Transport in Porous Media

Charges are present in the pore space of a porous medium in the form of ions. Since these ions represent solute species, their amounts are quantified by concentrations C_i, which are directly related to ionic charge, q_i, via charge number z_i. In other words, charge contribution of species i having concentration C_i (mol/m^3) is $q_i = z_i F C_i$, where $F = 96487$ C/mol, is Faraday's constant. Charge number is the charge per unit ion with appropriate sign, for example, $z_i = +1$ for Li^+ cation and $z_i = -2$ for SO_4^{2-} anion. Note that both C_i and q_i are intrinsic (phase) average quantities (i.e., defined per unit volume of the pore space). The corresponding ionic flux is quantified by Nernst–Planck equation [18], and the equivalent volume-averaged relation for transport in pore space is given by Eq. (3.1):

$$N_i = -D_i \frac{\varepsilon}{\tau} \nabla C_i - z_i F B_i \frac{\varepsilon}{\tau} C_i \nabla \phi_e + C_i \mathbf{u} \tag{3.1}$$

The presence of porosity/tortuosity ratio (ε/τ) accounts for the additional transport resistance in the pore network of a porous volume. Symbol $\mathbf{u}$ is the bulk-volume (superficial)-averaged velocity vector, which is often computed via

Darcy's equation. The Nernst–Planck expression describes the three components of ionic flux: (i) diffusion—flux in response to concentration gradient ∇C_i, (ii) migration—flux in response to gradient in electric potential $\nabla \phi_e$, and (iii) advection—flux due to solvent velocity field $\mathbf{u}$ [appearing in the same sequence on the right-hand side of Eq. (3.1)]. Here, B_i is ionic mobility and is related to ionic diffusivity via Stokes–Einstein relation $B_i = D_i/RT$ (R is universal gas constant and T is temperature). Using this flux description Eq. (3.1), the species balance statement for species i is written as follows:

$$\varepsilon \frac{\partial C_i}{\partial t} + \nabla \cdot (\mathbf{u} C_i) = \nabla \cdot \left(D_i \frac{\varepsilon}{\tau} \nabla C_i \right) + z_i F \nabla \cdot \left(\frac{D_i}{RT} \frac{\varepsilon}{\tau} C_i \nabla \phi_e \right) + r_i \qquad (3.2)$$

Equation 3.2 can be understood as an extension of the advection-diffusion equation of (neutral) species balance for charged species. Here, r_i denotes the net rate of production for ith ionic species. Equation (3.2) assumes that the pore phase geometry is time invariant.

The electric field in the solution phase ϕ_e is described by the statement of charge conservation in the solution phase Eq. (3.3), where the source terms J_j relate to jth electrochemical reaction. Equation (3.3) can be succinctly written in terms of total ionic flux $I = \sum_i z_i F N_i$ as expression (3.5). Here [in Eq. (3.3)], the first summation that appears as a coefficient to electric potential gradient is the effective ionic conductivity, κ^{eff}, of the solution phase (Eq. 3.4). The next summation represents diffusional current contribution due to each of the charged species. The electrochemical reaction source terms, J_j, are related to the species balance source terms, r_i via reaction stoichiometry. Also, note that the r_i also have additional contributions from chemical reactions.

$$\nabla \cdot \left(\left(\sum_i z_i^2 F^2 \frac{D_i}{RT} \frac{\varepsilon}{\tau} C_i \right) \nabla \phi_e \right) + \sum_i \left(\nabla \cdot \left(z_i F D_i \frac{\varepsilon}{\tau} \nabla C_i \right) \right) + \sum_j J_j = 0 \quad (3.3)$$

$$\kappa^{eff} = \sum_i z_i^2 F^2 \frac{D_i}{RT} \frac{\varepsilon}{\tau} C_i \qquad (3.4)$$

$$-\nabla \cdot I + \sum_j J_j = 0 \qquad (3.5)$$

Electrochemical reactions, in general, describe interconversion between solution and solid phase charge carriers (i.e., ions and electrons, respectively). Thence, the volumetric current source term $\sum_j J_j$ simultaneously appears in solution and solid phase charge conservation statements (expressions (3.3) and (3.6), respectively) with opposite signs.

$$\nabla \cdot \left(\sigma^{eff} \nabla \phi_s \right) = \sum_j J_j \qquad (3.6)$$

Here, ϕ_s is electric field potential in solid phase, and σ^{eff} is effective electronic conductivity of the solid phase. Both the electric potentials, ϕ_s and ϕ_e, jointly define the overpotential driving the electrochemical reactions. Note that if the porous medium is electronically insulating (e.g., separator used in batteries), the solid phase potential equation has a trivial solution and is not solved. Moreover, there are no electrochemical reactions in separator and the source terms vanish from Eq. (3.3). Note that the r_i could still be nonzero, given the contributions from chemical reactions.

Thus, the general description of charge transport in porous media requires a set of four governing equations, each describing a different physical phenomenon:

i. Species balance for charged species Eq. (3.2);
ii. Momentum equation for fluid phase (e.g., Darcy's law to obtain volume-averaged velocity field $\mathbf{u}$);
iii. Charge conservation in solution phase Eq. (3.3);
iv. Charge conservation in solid phase Eq. (3.6).

These governing equations are solved with an appropriate set of boundary conditions. Depending on physical operation, different boundary conditions manifest. For example, consider the interface between a film electrode and a separator (similar to lithium–separator interface in Li-ion half-cells). Electrochemical reaction (s) takes place at this boundary and appropriately acts as an influx for corresponding ions. Mathematically,

$$-D_i \frac{\varepsilon}{\tau} \nabla C_i \cdot \hat{n} = I_i / F, \mathbf{u} \cdot \hat{n} = 0, \nabla \phi_e \cdot \hat{n} = 0, \text{ and } \nabla \phi_s \cdot \hat{n} = 0,$$

where I_i is electrochemical rate of generation of species i in the solution phase, $\hat{n}$ is unit normal pointing toward the separator, and $\mathbf{u} \cdot \hat{n} = 0$ expresses a boundary impermeable to the solution phase. As another example, consider an electrode–current collector interface. In a porous electrode, total current has two components (i) ionic and (ii) electronic (present in pore and solid phases, respectively). At the electrode–current collector interface, total current becomes completely electronic in nature, thus giving rise to the boundary conditions:

$$\nabla C_i \cdot \hat{n} = 0, \mathbf{u} \cdot \hat{n} = 0, \nabla \phi_e \cdot \hat{n} = 0, \text{ and } -\sigma^{eff} \nabla \phi_s \cdot \hat{n} = I_{app},$$

where I_{app} is total electronic current and $\hat{n}$ is local normal pointing from electrode to current collector direction. Let us discuss an example with flow. At the inlet of a redox flow battery electrode, the electrolyte is a sufficiently homogenized solution and is being pumped in at a fixed flow rate. This leads to the boundary conditions: $C_i = C_{i0}, \mathbf{u} \cdot \hat{n} = U_0, \nabla \phi_e \cdot \hat{n} = 0$, and $\nabla \phi_s \cdot \hat{n} = 0$. Here, C_{i0} is concentration of ionic species i in the reservoir, U_0 is inlet velocity, while the two potential boundary conditions describe that the incoming electrolyte is charge neutral and does not introduce any ionic or electronic current. $\hat{n}$ is unit normal at the inlet pointing toward the porous electrode of the flow battery.

3.2.1 Special Considerations

Many assumptions of the charge transport behavior described above are not always valid and require extension or generalization of these concepts. Three such situations are briefly described hereafter.

i. Variable porosity (pore space) field

The porous medium can have spatial variation of the effective properties (i.e., porosity, ε, tortuosity, τ, effective conductivity, σ^{eff}, etc.). The set of Eqs. (3.2), (3.3), and (3.6) already account for this spatial variation, as the effective properties are inside the spatial derivative. Keep in mind that the Darcy's equation would also equivalently have a spatially varying permeability field. These equations require further generalization when the pore network also evolves in time (e.g., due to precipitation/ dissolution reactions). For such events, the porosity appearing along with the temporal derivative in Eq. (3.2) has to be placed inside the time derivative operator, and equivalently the solution procedure has to be modified.

ii. Concentrated solution theory

One fundamental assumption of the Nernst–Planck equation is that the different solutes present in the solution do not affect each other. This is only valid when the solution is sufficiently dilute (usually millimolar, mM, concentrations) such that the hydrodynamic boundary layers of different ions do not overlap and they do not influence the dynamics of each other. As solute concentration increases (of the order of one molar, M, and above), these interspecies interactions have to be modeled as well. In such situation, the Nernst–Planck expression has to be replaced with more rigorous Stefan–Maxwell–Onsager relation relation [19]. This only changes the constitutive model, while the fundamental physics of species balance and charge conservation is still equally valid. This revised formulation is often referred to as the concentrated solution theory. Such a formulation explains the experimentally observed non-monotonic ionic conductivity trends [the dilute solution theory says that the ionic conductivity increases linearly with solute concentration, i.e., Eq. (3.4)].

iii. Electric double layer

Another inherent assumption in the above description is that of charge neutrality. It is assumed that the electrolyte solution is locally charge neutral, i.e., $\sum_i z_i C_i = 0$ (note that the solution is always globally charge neutral). The bulk of electrolyte exhibits local charge neutrality, but under certain circumstances the solid–pore phase interface can have a local charge distribution and the description has to be appropriately amended. For example, when a Li-ion cell is operated with a time-varying current field (i.e., during electrochemical impedance spectroscopy, EIS, measurements), these double layer effects become important. Physically, this manifests as an additional volumetric current source term (for a porous media description upon volume-averaging). This capacitive current is often referred to as a non-faradic contribution to distinguish it from the faradic (electrochemical) current. As long as current changes are negligible, electric double layer effects do not have to be incorporated, e.g., usual operation of batteries (constant current, constant voltage, etc.) and the cyclic voltammetry, CV, studies to name a few.

3.3   Mesoscale Models in Porous Media

Mesoscale modeling techniques for multiphase flows in porous media can be broadly classified into rule-based and first principle-based models. Rule-based models involve pore network (PN) extraction from the porous medium and solution of two-phase transport on this idealized network for prediction of transport properties [20]. The accuracy of this model is predicated on the robustness of the network extraction algorithm with majority PN models incorporating simplistic description of the porous media as spherical pores connected by cylindrical throats.

Alternatively, the first principle-based models solve the fundamental Navier–Stokes (NS) equations governing fluid flow, thereby resulting in highly accurate results, albeit with higher computational cost. Fine scale computational fluid dynamics (CFD) models [21, 22], molecular dynamics (MD) models [23], lattice Boltzmann models (LBM), and direct numerical simulation (DNS) [24] form the major candidates for first principle-based models.

Fine scale CFD models involve discretization of the NS equations on a computational grid using finite difference, finite element, or finite volume framework. CFD models such as front capturing and front tracking have emerged as techniques to simulate multiphase flow but are invariably limited by the complexity of application of boundary conditions on randomly shaped peripheries. Alternatively, the MD approach involves modeling the fluid as an ensemble of molecules and involves tracking the position and movement of each molecule in the ensemble using accurate inter-molecular force descriptors. For application to macroscopic flows in porous media, statistically significant results require the number of molecules in the ensemble to approach prohibitively large numbers, thereby causing the MD approach to become computationally redundant. The formulations of LBM and DNS models circumvent the aforementioned issues and have been widely utilized in simulating two-phase flow in PEFC. Hence, we focus our attention on these two models for the rest of the chapter.

3.4 Microstructure Generation

LBM and DNS models require an exhaustive description of the porous microstructure in the form of 3-D volume data as input. Experimental imaging techniques and stochastic simulation methods are the two routes to generate this dataset. Recent advancements in noninvasive experimental imaging techniques, such as X-Ray micro-tomography [25, 26], have been successfully utilized in generating 3-D microstructures of porous media. Alternatively, stochastic reconstruction techniques harness precise statistical information (like porosity, two-point correlation function) about the porous specimen from a 2-D image to create virtual realizations of the 3-D microstructure [27, 28]. Stochastic reconstruction-based microstructure generation has been utilized extensively to create PEFC CL [29, 30] and GDL [31, 32] microstructures. Figure 3.3 shows a stochastically reconstructed CL microstructure with 60% nominal porosity and 10 μm thickness generated using an input TEM (transmission electron microscope) micrograph. The solid and pore phase fraction variation along the thickness direction is also delineated [29]. The reconstructed microstructure of a carbon paper GDL with porosity 72% and thickness 18 μm along with the pore size distribution is presented in Fig. 3.4 [33].

3.5 Lattice Boltzmann Modeling

The lattice Boltzmann method (LBM), having its origin in classical statistical physics, is a mesoscopic approach based on simplified kinetic equations and has shown tremendous success in simulating fluid flow applications involving interfacial dynamics and complex geometries, e.g., multiphase/multicomponent flows in porous medium, within a computationally tractable framework. The origins of the LBM approach lie in the lattice gas automata (LGA) method involving discrete space, time, and particle velocities [34]. LB models the fluid as a collection of fictitious particles residing on a discrete lattice mesh with propagation and collision of the particles governed by a velocity distribution function. In this regard, it is inherently similar to MD simulations requiring the tracking of individual fluid particles, albeit fictitious. The upshot of the LB method lies in its algorithmic design which makes it extremely amenable to high-performance computing using massively parallel architectures. Parallel execution is an inherent component of the LB approach and this enables the rapid execution of sophisticated algorithms for incorporation of complex physics, making it highly attractive for modeling two-phase flows in porous media.

Another important aspect of the LB approach is its ability to bridge multiple length and timescales (see Fig. 3.5). Numerical schemes implemented in the LB model incorporate simplified kinetic models that can capture essential microscopic and mesoscopic flow physics, with the averaged quantities simultaneously satisfying the macroscopic Navier–Stokes equations governing fluid flow.

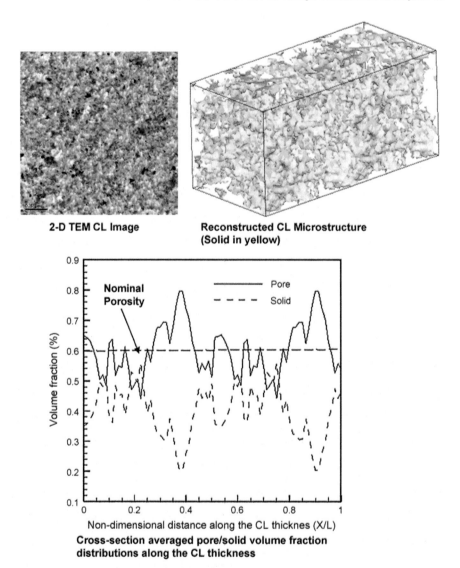

Fig. 3.3 Reconstructed catalyst layer microstructure along with pore and electrolyte phase volume fractions distribution [15]

Numerous two-phase LB models studying multiphase and multicomponent flows are readily available in the literature. Initial efforts led to the development of a two-component lattice gas model by Gustensen et al. [35] using red and blue colored particles to represent two different fluids. Shan and Chen [36, 37] implemented an inter-particle interaction potential in the LB framework capable of simulating phase separation, hereafter known as the S-C model. Further developments involved usage of free energy approach based on non-equilibrium dynamics

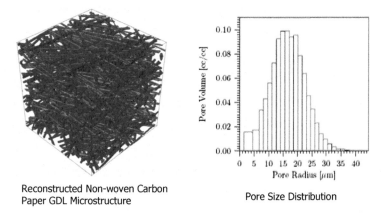

Reconstructed Non-woven Carbon
Paper GDL Microstructure

Pore Size Distribution

Fig. 3.4 Reconstructed nonwoven carbon paper GDL microstructure along with the evaluated structural properties [15]

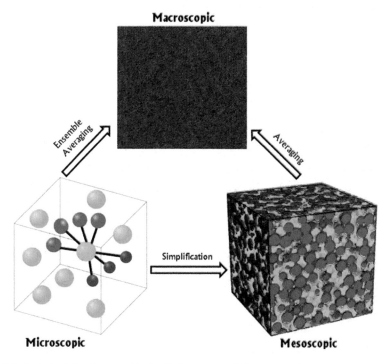

Fig. 3.5 Scale-bridging feature of lattice Boltzmann modeling approach

by Swift et al. [38, 39] and employing two sets of probability distribution function aimed at simulating pressure, velocity fields and interface capturing, respectively, by He et al. [40, 41], with both methods incorporating more physical realism in the models. However, they were afflicted with numerical instabilities rendering limited applicability for multiphase flows. The S-C model was revamped considerably and is now widely used for simulating multiphase flows in porous media [42, 43]. Major advantages of the S-C model include simpler implementation of complex boundary conditions present in morphologically diverse porous structures, flexible handling of fluid phases with differing densities, viscosities and wettabilities, and versatility in incorporating varying equations of state. Consequently, LBM-based two-phase transport in PEFC is mainly investigated using the S-C model.

3.5.1 Methodology

The S-C model considers a fluid mixture comprising of k different components distributed on a lattice. Each of these components is represented by k distribution functions which satisfy the spatiotemporal evolution equation. Particles at neighboring sites interact through a set of non-local interaction potentials satisfying non-ideal gas equations of state. The collision and propagation step for the kth component can then be written as:

$$f_i^k(\mathbf{x} + \mathbf{e}_i \delta_t, t + \delta_t) - f_i^k(\mathbf{x}, t) = -\frac{f_i^k(\mathbf{x}, t) - f_i^{k(eq)}(\mathbf{x}, t)}{\tau_k} \qquad (3.7)$$

The meaning of the individual terms is elaborated here. $f_i^k(\mathbf{x}, t)$ is the number density distribution function for the kth component in the ith velocity direction at position $\mathbf{x}$ and time t, and δ_t is the incremental time step. The complete right-hand side term represents the collision term based on the BKG (Bhatnagar–Gross–Krook) single relaxation time approximation while τ_k is the mean relaxation time for the kth component in lattice unit which determines its fluid viscosity. $f_i^{k(eq)}(\mathbf{x}, t)$ is the corresponding equilibrium distribution and has a defined functional form for a particular lattice configuration. The discrete form of Eq. (3.7) is also referred to as the LBGK equation. A typical three-dimensional 19-velocity lattice (D3Q19, where D is the dimension and Q is the number of velocity directions), with the velocity directions, is shown schematically in Fig. 3.6. The nomenclature implies that the particles entering/leaving at each lattice node can move along eighteen different directions or rest at the node. Alternative models that are used include D3Q15 and D3Q27 models but the D3Q19 model is favored and outperforms the other two configurations in terms of computational reliability and efficiency. Different fluid phases exhibit varying properties like wettability on solid surface and phase separation characteristics. The incorporation of these variations as well as presence of body force requires modifications in the velocity used to calculate the equilibrium distribution function.

Fig. 3.6 Schematic of
D3Q19 lattice structure [44]

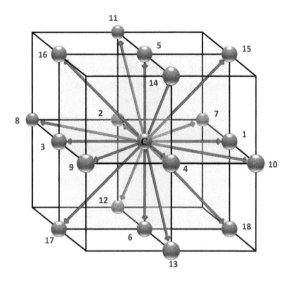

An extra component-specific velocity due to inter-particle interaction is added on
top of a common velocity for each component, the magnitude of which depends on
the total force $\mathbf{F}_k$. Correspondingly, inter-particle interaction is realized through the
total force, $\mathbf{F}_k$, acting on the kth component, including fluid/fluid interaction,
fluid/solid interaction, and external force. Fluid/ fluid interaction in the D3Q19
model takes into account the nearest and next nearest neighbors. Interaction
between fluid and wall (fluid/solid interaction) is incorporated by considering the
wall as a separate phase with constant number density while constant body forces
(like gravity) are easily computed.

The continuity and NS momentum equations can be obtained for the fluid
mixture as a single fluid using Chapman–Enskog expansion procedure in the nearly
incompressible limit and is given in Eq. (3.8). It is worthwhile to note that the S-C
model does not conserve local momentum at each lattice site but the total
momentum of the system is conserved.

$$\left.\begin{array}{l} \frac{\partial \rho}{\partial t} + \nabla \cdot (\rho \mathbf{u}) = 0 \\ \rho \left[\frac{\partial \mathbf{u}}{\partial t} + (\mathbf{u} \cdot \nabla) \mathbf{u} \right] = -\nabla p + \nabla \cdot \left[\rho v (\nabla \mathbf{u} + \mathbf{u} \nabla) \right] + \rho \mathbf{g} \end{array}\right\} \quad (3.8)$$

Here, the total density and velocity of the fluid mixture are given, respectively,
by:

$$\left.\begin{array}{l} \rho = \sum_k \rho_k \\ \rho \mathbf{u} = \sum_k \rho_k \mathbf{u}_k + \frac{1}{2} \sum_k \mathbf{F}_k \end{array}\right\} \quad (3.9)$$

with a non-ideal gas equation of state for pressure.

LB model implementation requires the determination of several physical parameters like fluid/fluid and fluid/solid interaction parameters. For this purpose, two-phase LB model calibration is performed using benchmark numerical computations. Surface tension force manifests from fluid/fluid interaction while fluid/solid interaction triggers the wall adhesion force. Fluid/fluid interaction parameter is evaluated using the spherical *bubble test* in absence of solid phase with Young-Laplace law used to determine the veracity of the simulations. *Static droplet test* is performed to deduce the fluid/solid interaction parameter using contact angle computations. Detailed analysis of these numerical experiments are presented in Mukherjee et al. [44].

3.5.2 Representative Highlights

The evaluation of non-dimensional numbers in two-phase fluid flow through porous PEFC electrodes incorporating surface tension, gravity, viscous, and inertial forces exhibits the dominance of surface tension forces as compared to the gravity, viscous, and inertial force. Typical PEFC electrode microstructures exhibit significantly small pore sizes (e.g., ~ 0.1 μm in the CL and ~ 20 μm in the GDL) and meager flow rates. The low magnitudes of Reynolds number $\left(Re = \rho UD/\mu \sim 10^{-4}\right)$, Capillary number $\left(Ca = \mu U/\sigma \sim 10^{-6}\right)$, Bond number $\left(Bo = g\Delta\rho D^2/\sigma \sim 10^{-10}\right)$ and Weber number $\left(We = \rho U^2 D/\sigma \sim 10^{-10}\right)$ in both the GDL and CL illustrate the strong influence of surface tension force for air–water two-phase flow in PEFC electrode. Here, ρ, U and μ are the non-wetting phase density, velocity, and dynamic viscosity, respectively; D is the representative pore size; σ is the surface tension, and g is the gravitational acceleration.

The LB model has been deployed for investigating liquid water transport and two-phase dynamics through the reconstructed PEFC CL and GDL microstructures utilizing appropriate density and viscosity ratios [44]. Primary drainage simulation is performed to model a quasi-static displacement experiment involving immiscible, two-phase transport through the CL and the GDL. The setup involves the addition of a non-wetting phase (NWP) and a wetting phase (WP) reservoir at the front and back end of the porous structure, respectively. The added reservoirs are oriented in the through plane (i.e., thickness) direction and are composed of void space. The hydrophobic characteristics of the CL and GDL imply that liquid water is the NWP and air is the WP. Dirichlet boundary conditions (BC) are imposed for pressure at the first layer of the NWP reservoir and the last layer of the WP reservoir while periodic boundary conditions are applied in the span-wise directions. Dirichlet pressure BC transform to fixed densities within the LB framework. Its implementation involves computation of equilibrium distribution functions using zero velocity and specified density at the reservoirs. No-slip boundary condition is applied at the walls utilizing a particle distribution function bounce back scheme. At the start of the simulation, equal NWP and WP reservoir pressures are maintained which gives a net zero capillary pressure. Steady initiation of the primary drainage simulation is conducted by incrementally decreasing the WP reservoir pressure while keeping the NWP reservoir pressure at the fixed initial value. Pressure gradient-induced liquid water flow is observed in the initially air-saturated CL and GDL through displacement of the air. The quasi-static displacement simulation provides an understanding of the nature of liquid water transport navigation through the pore morphology of CL and GDL and the ensuing capillarity effects.

Figure 3.7 shows the steady-state advancing liquid waterfronts inside CL microstructure in response to capillary pressure increments in the primary drainage simulation. The CL microstructure used exhibits hydrophobic wetting characteristics with a static contact angle of 100°. The initially air-saturated microstructure (wetting fluid) shows larger liquid water infiltration with higher capillary pressure. Capillary fingering regime is observed at low capillary pressures whereby the liquid

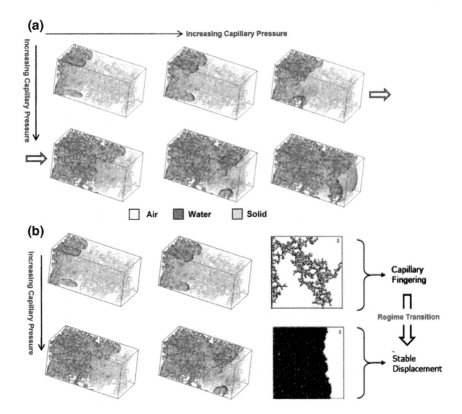

Fig. 3.7 Advancing liquid waterfront with increasing capillary pressure through the initially air-saturated reconstructed CL microstructure from the primary drainage simulation using LBM [44]

water saturation fronts exhibits finger-like patterns. Surface tension-driven capillary force drives the penetration of liquid water phase into the resident air wetting phase region in the shape of fingers. With increasing capillary pressure, the capillary fingering regime transitions to the stable displacement regime owing to the merging of several penetrating saturation fronts. This study points to an interesting conclusion; even for extremely low capillary number flows inside the CL, the advancing liquid waterfront metamorphoses from a fingering structure to a somewhat flat structure as the capillary pressure rises.

Primary drainage simulations using two-phase LBM are also performed by Mukherjee et al. [44] for the carbon paper GDL exhibiting higher hydrophobicity as compared to the CL investigated above, with a static contact angle of 140°. Figure 3.8 displays the liquid water distribution as well as the intrusion pattern with increasing capillary pressure in the initially air-saturated GDL. Owing to the stronger hydrophobicity and larger pore size, liquid waterfront incursion and droplet formation ensue only at some preferential locations for low capillary pressure. As the capillary pressure increases, formation and penetration of several waterfront is observed due to capillarity. The multiple fronts coalesce and form two major chunks which advance in the less tortuous in-plane direction. Beyond a threshold value of capillary pressure, one of the fronts reaches the WP air reservoir, the physical equivalent of the GDL/air interface, at a preferred location which is termed the bubble point. These simulations reveal the strength of two-phase LBM models which can capture intricate liquid water dynamics including droplet formation, coalescence, and front propagation through the hydrophobic fibrous GDL structure.

The extensive applicability of the LBM two-phase model is further demonstrated for carbon paper GDL structures characterized by *mixed wettability*. Fresh GDL exhibits fully hydrophobic characteristics minimizing liquid water flooding. Protracted PEFC operation can, however, cause the GDL to lose its hydrophobicity, thereby increasing its susceptibility to water flooding. Experimental data suggests oxidation of GDL surface carbon atoms to hydrophilic carboxyl groups/phenol as a result of long-term exposure to strong oxidative conditions in the cathode during PEFC operation. Consequently, Mukherjee et al. [44] investigated the repercussions of reduced GDL hydrophobicity on the flooding behavior. *Mixed wettability* is realized using randomly dispersed hydrophilic and hydrophobic pores inside the initially air-saturated GDL microstructure. The static contact angles exhibited by the hydrophilic and hydrophobic pores are taken to be 80° and 140°, respectively, which are characteristic of aged GDL structures. Figure 3.9 shows the invading liquid waterfront profiles from the primary drainage simulations for varying capillary pressures. The presence of both hydrophobic and hydrophilic pores results in simultaneous water droplet and film formation at low capillary pressures. Further increment in capillary pressure leads to merging of the liquid water film fronts which assist in front movement. In contrast to fully hydrophobic GDL, liquid film formation dominates front advancement and GDL anisotropy is rendered ineffective in its ability to assist forking and in-plane motion of the front. Liquid water slug formation is observed as bubble point is approached as opposed to finger like structures leading to higher saturation levels and enhanced flooding.

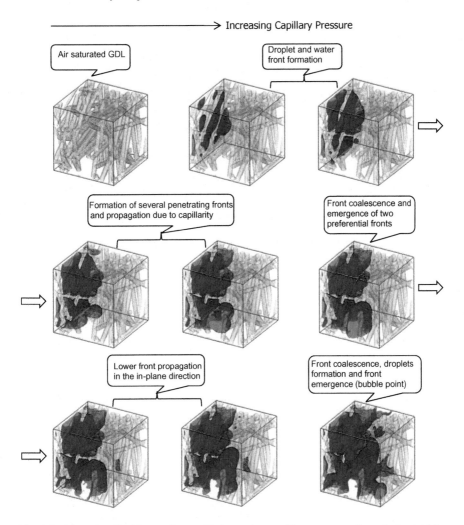

Fig. 3.8 Advancing liquid waterfront with increasing capillary pressure through the initially air-saturated reconstructed GDL microstructure from the primary drainage simulation using LBM [44]

3.6 Electrochemistry-Coupled Direct Numerical Simulation

The conventional approach to modeling porous electrodes in fuel cells utilizes the macrohomogeneous models incorporating volume-averaging of characteristic phase properties and variables over a representative elementary volume (REV) containing statistically significant number density of particles. In this method, the intricate microscopic details of the pore structure are implicitly resolved and homogenization over the electrode volume is performed to obtain averaged geometric and transport

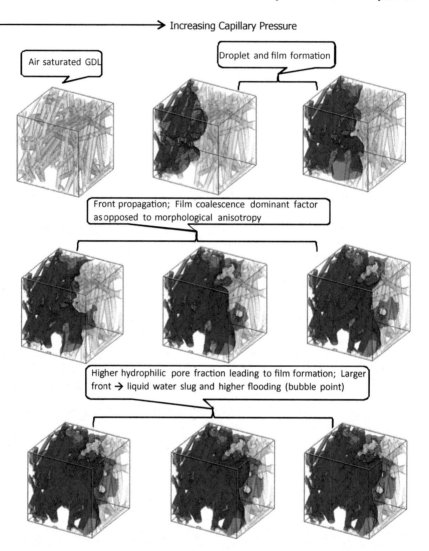

Fig. 3.9 Advancing liquid waterfront with increasing capillary pressure through the mixed wet GDL microstructure with 50% hydrophilic pores from the displacement simulation using LBM [44]

properties like porosity, specific surface area, tortuosity, effective diffusivity, effective conductivity. Typically, empirical correlations are used to describe the effective properties as a function of porosity and tortuosity of the porous medium. The use of empirical correlations exacerbates the inaccuracy of the macrohomogeneous models. In this regard, direct numerical simulations (DNS) performed over the fully resolved porous microstructure is an accurate, albeit, computationally expensive tool to study the influence of the microstructure on the species and charge transport processes occurring inside the porous media. The DNS methodology also provides a tool to improve the accuracy of the empirical correlations used in the macrohomogeneous models, giving an accurate estimate of the magnitude of Bruggeman factors appearing in the correlations. In this section, we describe the salient features of DNS modeling and its usage in characterizing transport inside the catalyst layer of PEFC electrodes.

3.6.1   Methodology

Reaction kinetics, species, and charge transport characterize the phenomenological processes occurring inside the porous electrode of PEFC. Correspondingly, the phenomenological events considered in the current DNS model [30] are the following:

1. Charge transfer kinetics of oxygen reduction reaction (ORR) occurring at the electrochemically active sites characterized by the triple phase boundary;
2. Species transport (oxygen and water vapor) though the pore phase;
3. Charge transport (H^+) through the electrolyte phase.

A few assumptions are used to simplify the DNS model development which are detailed below:

(a) Isothermal, steady-state operation of the PEFC;
(b) Negligible oxygen diffusion resistance though polymer electrolyte film to Pt sites;
(c) Uniform electronic potential in the CL;
(d) Uniform liquid water content.

The above assumptions are justified in their usage owing to the small thickness and high electronic conductivity of the catalyst layer.

Single domain approach is utilized to develop valid governing differential equations for all the phases, precluding the need for specification of boundary conditions at phase interfaces. Tafel approximation is used to describe the sluggish ORR kinetics and has the following form:

$$j = -i_0 \left(\frac{c_{O_2}}{c_{O_2,ref}} \right) \exp\left(\frac{-\alpha_c F}{RT} \eta \right) \tag{3.10}$$

Here, i_0 is the exchange current density, c_{O_2} and $c_{O_2,ref}$ refer to local and reference oxygen concentration, respectively, α_c is the cathodic transfer coefficient for ORR, F is the Faraday's constant, R is the universal gas constant, and T is the cell operating temperature. A representative value of exchange current density value for ORR in fuel cells is 50 nA/cm^2. The overpotential, η, is defined as:

$$\eta = \varphi_s - \varphi_e - U_0 \tag{3.11}$$

Here, φ_s and φ_e denote the solid and electrolyte phase potentials at the reaction sites, respectively. The open circuit potential of the cathode at the reference temperature is designated U_0.

Proton (H^+) charge transport and species transport of oxygen and water vapor are governed by the following Poisson conservation equations.

$$\left. \begin{array}{l} \nabla \cdot (\kappa_e \nabla \varphi_e) + S_{\varphi_e} = 0 \\ \nabla \cdot (D^g_{O_2} \nabla c_{O_2}) + S_{O_2} = 0 \\ \nabla \cdot (D^g_{H_2O} \nabla c_{H_2O}) + S_{H_2O} = 0 \end{array} \right\} \tag{3.12}$$

Here, the source terms for proton, oxygen, and water vapor denoted by S_{φ_e}, S_{O_2}, and S_{H_2O} arise due to the electrochemical oxygen reduction reaction. φ_e is the electrolyte phase potential, c_{O_2} and c_{H_2O} are the oxygen and water vapor concentration inside the pore phase, respectively. The transport parameters are denoted by κ_e, $D^g_{O_2}$, and $D^g_{H_2O}$ denoting protonic conductivity, oxygen, and water vapor diffusivity, respectively. Exhaustive details about the DNS model for pore-scale description of species and charge transport in the CL microstructure along with its capability in discerning the compositional influence on the CL performance as well as local overpotential and reaction current distributions are available in the literature [29, 30]. It is important to note that the mass transport resistance due to liquid water transport is not considered in the current DNS model.

3.6.2 Representative Results

DNS computations have been performed for both monolayer [17, 29] and multilayer CL [30] microstructures. An important application of the computed DNS data is the evaluation of Bruggeman correlation required for macrohomogeneous models. The macrohomogeneous models, as introduced earlier briefly, solve homogenized transport equations using effective geometric and transport properties [45, 46]. Effective property correlations usually involve a Bruggeman correction factor, ζ, and are given by:

$$\Gamma_k^{eff} = \Gamma_k \cdot \varepsilon_k^{\zeta} \tag{3.13}$$

Here, Γ_k^{eff} is the effective transport property and ε_k is the volume fraction of the transport medium k. Similar to the Bruggeman factor, another measure of the transport resistance often encountered in porous media literature is the tortuosity, τ, and can be expressed as:

$$\Gamma_k^{eff} = \Gamma_k \cdot \frac{\varepsilon_k}{\tau_k} \tag{3.14}$$

Physically, tortuosity is defined as the ratio of the actual distance traversed by the species to the shortest distance between the species start and end points. It is a measure of the convolutedness of the medium transport path. The computation of either Bruggeman factor or tortuosity facilitates the use of computationally inexpensive 1-D macrohomogeneous models. Consequently, determination of both the Bruggeman correction factor and tortuosity is attempted once the results of the DNS computations are available.

Figure 3.10 shows the results of the 3-D DNS computations alongside the 1-D macrohomogeneous predictions on the monolayer CL microstructure [29]. The cross-sectional averaged reaction current density is plotted as a function of CL thickness for both models. Different Bruggeman factors are explored to get the best fit for the macrohomogeneous results with the DNS data. Best agreement between the two models is exhibited for Bruggeman factor of 3.5. The corresponding tortuosity values are computed to be 4 and 18 for pore phase and electrolyte phase tortuosity representative of oxygen and proton diffusion resistance. Higher proton path tortuosity is exhibited owing to the lower volume fraction of electrolyte compared to the pore.

Another significant observation is the steep decrease in reaction current density in the vicinity of the membrane–catalyst layer interface along the thickness of the catalyst layer. This can also be attributed to deficient ionomer conductivity resulting from low electrolyte phase volume fraction throughout the CL.

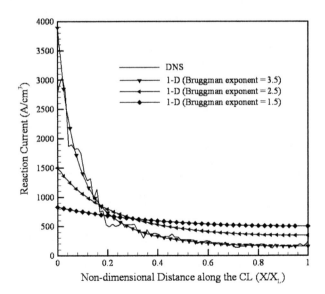

Fig. 3.10 Comparison between the cross-sectional averaged reaction current distributions across the thickness of the CCM CL from the DNS and 1-D macrohomogeneous models [29]

It is evident that disparate and competing transport mechanisms prevalent within the cathode catalyst layer require an optimal balance among the constituent phases in order to achieve the best cell performances. For example, an increase in Nafion® content improves proton conductivity while reducing the available pore space for oxidant transport resulting in significant decline in the gas phase diffusivity. Increase in platinum (Pt) loading, on the other hand, enhances the electrochemical reaction rate, but raises the PEFC cost exorbitantly. Another important factor affecting the PEFC performance is the involvement of water transport in the cathode CL, via water production due to the ORR as well as migration from the anode side by electroosmotic drag. Good proton conductivity requires hydration of the electrolyte phase which is again strongly dependent on the cell operating conditions in terms of the inlet relative humidity as well as the cell operating temperature. Conversely, high water content can clog the pores and choke the reaction sites leading to cell shutdown. It is, therefore, discernible that achieving an enhanced cell performance warrants detailed understanding of the dependence of the cathode CL performance on its composition as well as on the cell operating conditions. The following section details the investigation into the effects of cathode CL composition and cell operating conditions on its performance.

The aforementioned DNS model was deployed to study the influence of structural variation of a bilayer CL on the underlying transport and performance [30]. The design of the bilayer CL incorporates physical collocation of two catalyst-coated membrane (CCL) layers, designated **A** and **B**, each having a thickness around 5 μm and equal Pt loadings. The two layers differ in the ionomer to carbon (I/C) weight ratios, 0.417 for **A** and 0.667 for **B**. Cathode structure variation is accomplished using **A/B** or **B/A** CL arrangement in between membrane and the GDL. Figure 3.11 illustrates these two different composite CL designs. Pore and electrolyte volume fractions, ε_{CL} and ε_{Nafion}, are computed for each of the CL layers using the CL dimensions, Pt loading, ionomer to carbon and carbon to Pt weight ratios with the values given below:

$$\begin{aligned}\varepsilon_{CL/A} &= 0.56, \varepsilon_{Nafion/A} = 0.12 \\ \varepsilon_{CL/B} &= 0.48, \varepsilon_{Nafion/B} = 0.2\end{aligned} \tag{3.15}$$

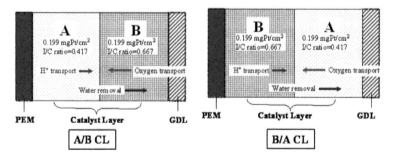

Fig. 3.11 Schematic representation of the bilayer CLs [30]

The computed volume fractions for pore, electrolyte, and electronic phases are used to reconstruct the **A/B** CL and **B/A** CL microstructures using stochastic generation method which are then provided as input to the DNS model. The results obtained from the full-blown DNS simulations are detailed here.

Baseline simulations were run for the **A/B** and **C/A** CLs with 100% relative humidity air, at 70 °C PEFC operation temperature, 200 kPa inlet pressure, and current density of 0.4 A/cm^2, which is on the lower side. Figure 3.12 shows the distribution of the cross-sectional averaged reaction current density and cathode overpotential along the thickness direction. Protonic resistance and oxygen transport are the two competing factors impacting catalyst layer performance. Charge transport resistance dominates performance at low current densities while hindered oxygen transport becomes significant at high current densities. Protonic resistance reduces with higher ionomer volume fraction, while oxygen transport is favored by higher pore volume fraction. The **B/A** CL is characterized by high ionomer fraction close to membrane–CL and large pore fraction close to CL–GDL with reverse trends for **A/B** CL. Thus, the **B/A** CL exhibits lower cathodic overpotential near the membrane–CL interface, with the larger ionomer content aiding protonic transport. An extended reaction zone is also seen for the **B/A** CL, evidenced by the reaction current distribution. Furthermore, higher pore volume fraction near the CL–GDL interface for **B/A** CL facilitates robust oxygen transport. In contrast, higher protonic resistance near membrane–CL interface for **A/B** CL necessitates larger magnitude of cathode overpotential and a decreased reaction zone. Smaller porosity near the CL–GDL interface further diminishes the performance of the **A/B** CL. Hence, **B/A** CL configuration dominates **A/B** CL in terms of performance. In conclusion, coupling the higher ionomer content close to membrane–CL interface with higher pore volume fraction in the vicinity of CL–GDL interface can improve the performance characteristics of the PEFC owing to enhanced proton as well as oxygen transport.

Figure 3.13 displays the 3-D reaction current contours for the baseline simulations as well as inlet relative humidity variation with **A/B** and **B/A** CL

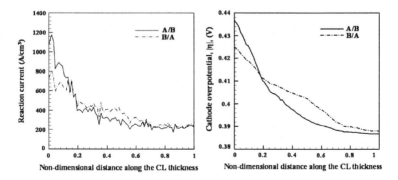

Fig. 3.12 Cross-sectional averaged reaction current and overpotential distributions along the thickness of the bilayer **A/B** and **B/A** CLs [30]

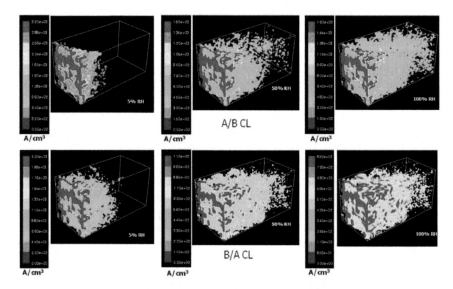

Fig. 3.13 3-D reaction current distribution contours for different inlet humidity conditions at 70 °C and 0.4 A/cm^2 for the **A/B** and **B/A** CLs [30]

microstructures. It is evident that high humidity operation is incumbent for good performance characteristics of the PEFC catalyst layer. Fully humidified operation results in protracted reaction current density distributions, thereby minimizing the overpotential losses. The results for 5% RH and 50% RH exhibit reaction current snap off which renders a significant portion of the CL virtually inactive. As explained earlier, the deterioration in performance is more pronounced for **A/B** CL as compared to **B/A** CL. An interesting observation is the apparent outperforming of the **B/A** CL by the **A/B** CL for lower humidity operation (see 50% RH) due to the slight extension of the ohmic control regime delaying the onset of mass transport limitations. Nevertheless, the suboptimal porosity near the CL–GDL interface compounded by the reaction zone shift toward membrane–CL interface in the **A/B** CL will inhibit oxygen transport resulting in inferior performance as compared to **B/A** CL. The results further emphasize the importance of the current DNS model in elucidating detailed pore-scale description of underlying transport through the CL microstructures.

The suitability of the DNS models can be verified by comparing the predicted polarization curves with actual experimental data as shown in Fig. 3.14. Electrochemical polarization curves were obtained for the full PEFC configuration consisting of a 5 cm^2 graphite cell fixture with identical anode and cathode single pass, serpentine flow fields. The simulation parameters were kept consistent with the experimental operating conditions of 70 °C, 200 kPa, and 100% RH at both anode and cathode sides with fixed flow rates of hydrogen and oxygen. Data processing of the experimental voltage vs current density results was performed to obtain the variation of cathode overpotential with current density against which the

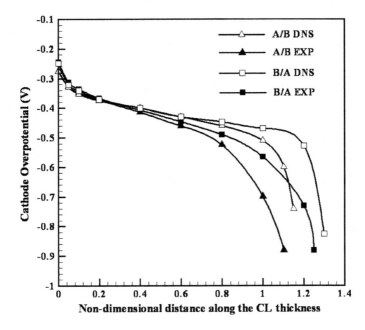

Fig. 3.14 Polarization curves for the bilayer **A/B** and **B/A** [30]

predicted results were compared. The results show reasonable concurrence in the kinetic and ohmic control regimes. However, the DNS calculations overpredict the ohmic control regime for both **A/B** and **B/A** CLs, owing to disregard of liquid water transport effects in the model. Lastly, both the experimental and DNS exhibit the superior performance of **B/A** over **A/B** CLs, which confirms the conclusion that higher ionomer content adjacent to membrane–CL interface and higher porosity near to CL–GDL interface, is indeed advantageous for bilayer CL performance.

3.7   Summary and Outlook

The multiphase multicomponent porous structure of PEFC constituents involve myriad kinetic and transport processes occurring at multiple length scales. Detailed analysis of the phenomenological processes occurring at the mesoscale and bridging it with the macroscale performance of the PEFC requires investigations into the pore scale with fully resolved 3-D electrode microstructures. In this chapter, the role of mesoscale interactions emanating from the underlying porous microstructure and transport phenomena in the PEFC is illustrated with examples taken from the lattice Boltzmann modeling of two-phase transport and electro-chemistry-coupled direct numerical simulations. The use of LB models in analyzing two-phase air–water flow inside the CL as well as the GDL of PEFCs is demonstrated. Finally, coupled transport and performance investigations using DNS models are shown and the link with the macrohomogeneous models is established. It is fairly evident that the information obtained from mesoscale modeling has great significance and is indispensable to understanding the performance of PEFCs. Further improvement of mesoscale models can help enhance the working of PEFC expanding its viability for energy applications.

Acknowledgements The authors of this chapter acknowledge Elsevier, Electrochemical Society, and Royal Society of Chemistry for the figures reproduced in this chapter from the referenced publications of their journals.

References

1. S. Gottesfeld, T.A. Zawodzinski, Polymer electrolyte fuel cells. Adv. Electrochem. Sci. Eng. **5**, 195–302 (1997)
2. S. Cleghorn et al., PEM fuel cells for transportation and stationary power generation applications. Int. J. Hydrogen Energy **22**(12), 1137–1144 (1997)
3. C.-Y. Wang, Fundamental models for fuel cell engineering. Chem. Rev. **104**(10), 4727–4766 (2004)
4. M. Hogarth, G. Hards, Direct methanol fuel cells. Platin. Met. Rev. **40**(4), 150–159 (1996)
5. R.M. Ormerod, Solid oxide fuel cells. Chem. Soc. Rev. **32**(1), 17–28 (2003)
6. Y. Wang et al., A review of polymer electrolyte membrane fuel cells: technology, applications, and needs on fundamental research. Appl. Energy **88**(4), 981–1007 (2011)
7. J.K. Nørskov et al., Origin of the overpotential for oxygen reduction at a fuel-cell cathode. J. Phys. Chem. B **108**(46), 17886–17892 (2004)
8. K. Broka, P. Ekdunge, Modeling the PEM fuel cell cathode. J. Appl. Electrochem. **27**(3), 281–289 (1997)
9. K. Tüber, D. Pócza, C. Hebling, Visualization of water buildup in the cathode of a transparent PEM fuel cell. J. Power Sources **124**(2), 403–414 (2003)
10. F. Zhang, X. Yang, C. Wang, Liquid water removal from a polymer electrolyte fuel cell. J. Electrochem. Soc. **153**(2), A225–A232 (2006)
11. U. Pasaogullari, C. Wang, Liquid water transport in gas diffusion layer of polymer electrolyte fuel cells. J. Electrochem. Soc. **151**(3), A399–A406 (2004)
12. U. Pasaogullari, C.-Y. Wang, Two-phase transport and the role of micro-porous layer in polymer electrolyte fuel cells. Electrochim. Acta **49**(25), 4359–4369 (2004)
13. U. Pasaogullari, C.-Y. Wang, K.S. Chen, Liquid water transport in polymer electrolyte fuel cells with multi-layer diffusion media, in *ASME 2004 International Mechanical Engineering Congress and Exposition* (American Society of Mechanical Engineers, 2004)
14. U. Pasaogullari et al., Anisotropic heat and water transport in a PEFC cathode gas diffusion layer. J. Electrochem. Soc. **154**(8), 823–834 (2007)
15. P.P. Mukherjee, Q. Kang, C.-Y. Wang, Pore-scale modeling of two-phase transport in polymer electrolyte fuel cells—progress and perspective. Energy Environ. Sci. **4**(2), 346–369 (2011)
16. S. Chen, G.D. Doolen, Lattice Boltzmann method for fluid flows. Annu. Rev. Fluid Mech. **30**(1), 329–364 (1998)
17. G. Wang, P.P. Mukherjee, C.-Y. Wang, Direct numerical simulation (DNS) modeling of PEFC electrodes: part I. Regular microstructure. Electrochim. Acta **51**(15), 3139–3150 (2006)
18. J. Newman, K.E. Thomas-Alyea, *Electrochemical Systems* (Wiley, Hoboken, 2012)

19. A.M. Bizeray, D.A. Howey, C.W. Monroe, Resolving a discrepancy in diffusion potentials, with a case study for Li-ion batteries. J. Electrochem. Soc. **163**(8), E223–E229 (2016)

20. M.J. Blunt, Flow in porous media—pore-network models and multiphase flow. Curr. Opin. Colloid Interface Sci. **6**(3), 197–207 (2001)

21. F.H. Harlow, J.E. Welch, Numerical calculation of time-dependent viscous incompressible flow of fluid with free surface. Phys. Fluids **8**(12), 2182–2189 (1965)

22. B.J. Daly, Numerical study of the effect of surface tension on interface instability. Phys. Fluids **12**(7), 1340–1354 (1969)

23. D.C. Rapaport et al., The art of molecular dynamics simulation. Comput. Phys. **10**(5), 456 (1996)

24. R. Ababou et al., Numerical simulation of three-dimensional saturated flow in randomly heterogeneous porous media. Transp. Porous Media **4**(6), 549–565 (1989)

25. P. Spanne et al., Synchrotron computed microtomography of porous media: topology and transports. Phys. Rev. Lett. **73**(14), 2001 (1994)

26. J. Fredrich, B. Menendez, T. Wong, Imaging the pore structure of geomaterials. Science **268** (5208), 276 (1995)

27. P. Adler, J.-F. Thovert, Real porous media: Local geometry and macroscopic properties. Appl. Mech. Rev. **51**(9), 537–585 (1998)

28. S. Torquato, Statistical description of microstructures. Annu. Rev. Mater. Res. **32**(1), 77–111 (2002)

29. P.P. Mukherjee, C.-Y. Wang, Stochastic microstructure reconstruction and direct numerical simulation of the PEFC catalyst layer. J. Electrochem. Soc. **153**(5), A840–A849 (2006)

30. P.P. Mukherjee, C.-Y. Wang, Direct numerical simulation modeling of bilayer cathode catalyst layers in polymer electrolyte fuel cells. J. Electrochem. Soc. **154**(11), B1121–B1131 (2007)

31. V.P. Schulz et al., Modeling of two-phase behavior in the gas diffusion medium of PEFCs via full morphology approach. J. Electrochem. Soc. **154**(4), B419–B426 (2007)

32. K. Schladitz et al., Design of acoustic trim based on geometric modeling and flow simulation for non-woven. Comput. Mater. Sci. **38**(1), 56–66 (2006)

33. S.H. Kim, H. Pitsch, Reconstruction and effective transport properties of the catalyst layer in PEM fuel cells. J. Electrochem. Soc. **156**(6), B673–B681 (2009)

34. U. Frisch, B. Hasslacher, Y. Pomeau, Lattice-gas automata for the Navier-Stokes Equation. Phys. Rev. Lett. **56**(14), 1505 (1986)

35. A.K. Gunstensen et al., Lattice Boltzmann model of immiscible fluids. Phys. Rev. A **43**(8), 4320 (1991)

36. X. Shan, H. Chen, Lattice Boltzmann model for simulating flows with multiple phases and components. Phys. Rev. E **47**(3), 1815 (1993)

37. X. Shan, H. Chen, Simulation of nonideal gases and liquid-gas phase transitions by the lattice Boltzmann Equation. Phys. Rev. E **49**(4), 2941 (1994)

38. M.R. Swift, W. Osborn, J. Yeomans, Lattice Boltzmann simulation of nonideal fluids. Phys. Rev. Lett. **75**(5), 830 (1995)

39. M.R. Swift et al., Lattice Boltzmann simulations of liquid-gas and binary fluid systems. Phys. Rev. E **54**(5), 5041 (1996)

40. X. He, S. Chen, R. Zhang, A lattice Boltzmann scheme for incompressible multiphase flow and its application in simulation of Rayleigh-Taylor instability. J. Comput. Phys. **152**(2), 642–663 (1999)

41. X. He et al., On the three-dimensional Rayleigh-Taylor instability. Phys. Fluids **11**(5), 1143–1152 (1999)

42. Q. Kang et al., Lattice Boltzmann simulation of chemical dissolution in porous media. Phys. Rev. E **65**(3), 036318 (2002)

43. Q. Kang, D. Zhang, S. Chen, Displacement of a two-dimensional immiscible droplet in a channel. Phys. Fluids **14**(9), 3203–3214 (2002)

44. P.P. Mukherjee, C.-Y. Wang, Q. Kang, Mesoscopic modeling of two-phase behavior and flooding phenomena in polymer electrolyte fuel cells. Electrochim. Acta **54**(27), 6861–6875 (2009)
45. T.E. Springer, T. Zawodzinski, S. Gottesfeld, Polymer electrolyte fuel cell model. J. Electrochem. Soc. **138**(8), 2334–2342 (1991)
46. M.L. Perry, J. Newman, E.J. Cairns, Mass Transport in Gas-Diffusion Electrodes: A Diagnostic Tool for Fuel-Cell Cathodes. J. Electrochem. Soc. **145**(1), 5–15 (1998)

Chapter 4
Porous Media Applications: Electrochemical Systems

Partha P. Mukherjee, Aashutosh Mistry, and Ankit Verma

4.1 Introduction

Lithium ion batteries (LIBs) have evolved as the favored electrochemical energy storage system for portable electronics in recent years. The commercial viability of LIBs is widening to large-scale applications, such as electric vehicles and grid storage, owing to their high energy and power density, good cycle life, and low self-discharge characteristics [1–6]. Li-ion cells represent an intercalation chemistry with Li^+ ions as the electroactive species. In the conventional design, Li^+ ions are stored in solid electrode active material via intercalation reaction. During operation, they shuttle back-and-forth between the two electrodes via liquid electrolyte (this electrolyte is made up of a suitable lithium salt, such as $LiPF_6$, dissolved in an appropriate organic electrolyte). The electrodes are made porous to increase surface area-to-volume ratio. This interfacial area between electrode and electrolyte represents the active interface at which electrochemical reactions take place.

A graphical illustration of a typical Li-ion unit cell is shown in Fig. 4.1. The cell contains porous separator sandwiched between two porous electrodes. Depending on the material content of these electrodes, one acts as anode (negative) and another as cathode (positive). Liquid electrolyte is filled in the void space. Consider the discharge operation of this cell. Li^+ ions stored in anode deintercalate and jump to electrolyte, generating electrons in the process. The generated electron travels from anode to cathode via external circuit (in response to electric field established in the solid phase). These electrons cannot move internally from anode to cathode due to the presence of electronically insulating separator. The Li^+ ions generated at anode–electrolyte interface due to deintercalation will move toward cathode due to a combined influence of concentration gradient-driven diffusion and electric field-induced migration. These ions will experience electric field established in electrolyte due to electrolyte phase potential gradient and eventually intercalate back at the cathode. The conventional choice for anode materials is graphite (LiC_6),

© Springer International Publishing AG 2018
M.K. Das et al., *Modeling Transport Phenomena in Porous Media with Applications*, Mechanical Engineering Series, https://doi.org/10.1007/978-3-319-69866-3_4

and a typical cathode is often a transitional metal oxide host (*LiMO*, MO = metal oxide). The corresponding cell reactions can be expressed as given in Eq. (4.1).

$$\text{Anode:} \quad LiC_6 \xrightleftharpoons[\text{charge}]{\text{discharge}} Li_{1-x}C + xLi^+ + xe^-$$
$$\text{Cathode:} \quad Li_{1-x}MO + xLi^+ + xe^- \xrightleftharpoons[\text{charge}]{\text{discharge}} LiMO$$
$$(4.1)$$

Efficient operation of a Li-ion cell relies on proper functioning of each of these transport processes. A finite rate operation leads to kinetic and transport limitations associated with these different processes to different extents. In turn, the cell performance differs from its theoretical (thermodynamic) behavior given different geometrical arrangements of various phases [7, 8] as well as operating conditions. Mathematical modeling has proved to be an attractive tool in investigating the

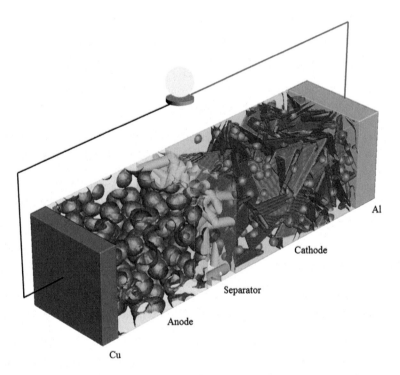

Fig. 4.1 Schematic diagram illustrating components of a typical Li-ion battery unit cell. Anode is made up of graphite particles, while cathode usually contains transition metal oxide as active material and appropriate amount of carbon to enhance electronic conductivity. Separator is an electronically insulating porous fibrous material (usually a polymer) to ensure that the cell is not internally shorted for electron transport. Only ionic transport is continuous internally. The electrodes are attached to metallic current collectors to ensure uniform current distribution at the ends

physics governing LIBs and helping us comprehend the behavior of these cells. A physical description of the intrinsic thermodynamic, transport, and kinetic processes within a computational framework helps delineate the practical limitations of the LIB and explore avenues for performance improvement. Modeling helps answer relevant questions such as:

1. How much is the internal resistance of the cell?
2. How much is the rate of heat generation for a given set of operational parameters?
3. How can we define working limits of the given cell, i.e., what are the values for safe current, voltage window of operation, etc.?

The present chapter explains established mathematical models for performance predictions of these cells, with suitable example simulations.

The rest of the chapter is arranged as follows. Thermodynamic, kinetic, and transport characteristics of LIB constituent materials, the building blocks for performance analyses, are detailed first in the following section. The governing equations pertaining to porous electrode theory [9–11] are presented in detail. Subsequently, performance results for isothermal LIB operation are discussed. Furthermore, thermal effects are also integrated within the aforementioned model to elucidate the non-isothermal operation of the cell. With the advent of composite electrodes, there have been arguments against the veracity of porous electrode approximation, and the use of models resolving intricate pore-scale details has been advocated. Different approaches falling into this realm have been elaborated upon in the final section.

The LIB models described in this chapter focus on performance analyses. Cell degradation over repeated usage has not been the focus of current discourse, and the readers are advised to refer to specialized research articles. Generally, deterioration of battery health can be attributed to two sets of events: (a) mechanical effects—repeated intercalation/deintercalation leads to stress cycling of the electrode materials causing damage and reduced efficacy of the intercalation process, and (b) chemical effects—highly reactive lithium metal reacts with electrolyte to produce insoluble and/or gaseous products, causing reduction in cyclable Li inventory. Major contributors in this category are Li dendrites and solid electrolyte interphase (SEI) formation. Moreover, cell failure can also result from anomalous operation, i.e., under abuse conditions like overcharge, external short, and thermal abuse [12–14]. Often, practical deterioration and/or failure is a combination of these diverse events.

4.2 Thermodynamic, Kinetic, and Transport Behavior of Li-Ion Battery Materials

Battery design subsumes selection of appropriate electrode active materials [15–17]. Electrode materials are chosen to have a large difference in open circuit potential between anode and cathode as well as ample Li storage capability. A materials road map for most of the materials of interest for present Li-ion battery usage is provided by Aurbach and coworkers [18] which reports average open circuit potential of the material along with their specific capacities. Here, open circuit potential (OCP) refers to the measured voltage with respect to a *Li* anode under open circuit conditions.

4.2.1 Open Circuit Potential

The open circuit potential for these electrode materials is a strong function of state of charge, i.e., amount of intercalated lithium. This is quantified as the ratio of *Li* amount present in the material to maximum intercable *Li* content for that material:

$$SOC = x = \frac{C_{Li}}{C_{Li}^{max}} \qquad (4.2)$$

Since open circuit potential is measured under thermodynamic equilibrium, Nernst equation [9, 19–21] should explain OCP variation as a function of intercalated *Li*. The corresponding mathematical formulation is

$$E = E^{\circ} - \frac{RT}{F} \ln\left(\frac{x}{1-x}\right) \qquad (4.3)$$

Nernstian behavior is plotted in Fig. 4.2 as a function of intercalated *Li* for metal oxide host MO. Note that the OCP approaches $+\infty(-\infty)$ as intercalation fraction go to $x \rightarrow 0$ (or 1). However, the Nernstian behavior does not match measured OCP profiles for almost all the *Li* intercalation host materials. This is evident from the OCP profiles of graphite and nickel manganese cobalt shown in Fig. 4.3. The profiles exhibit qualitative similarity with Nernstian behavior as $x \rightarrow 0$ (or 1). Deviation from Nernstian behavior is usually attributed to concentration-dependent activity coefficients [22–26].

4.2.2 Entropic Coefficient

The open circuit potential varies strongly with temperature, and this temperature dependence is captured in terms of entropic coefficient. Essentially, entropic coefficient is change in OCP with respect to changes in temperature. Like OCP, the entropic coefficients depend on the state of charge (Fig. 4.3 illustrate these functions for graphite and NMC). In addition to quantification of OCP at non-reference temperature $(T \neq 25\,^{\circ}\text{C})$, the entropic coefficients are also a measure of reversible heat generation magnitude. Let ΔH be the heat of reaction associated with a given intercalation reaction and ΔG be the corresponding available energy, i.e., amount of heat-to-work conversion. The difference is related to entropy change via

Fig. 4.2 Open circuit potential of a *Li* intercalated host Li_xMO as predicted by Nernst equation, E. Here, $E°$ refers to the OCP of the same under standard conditions of concentration and temperature

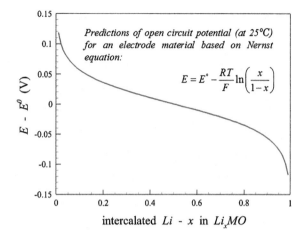

Predictions of open circuit potential (at 25°C) for an electrode material based on Nernst equation:

$$E = E° - \frac{RT}{F}\ln\left(\frac{x}{1-x}\right)$$

Fig. 4.3 Thermodynamic behavior of state-of-the-art electrode materials (**a**) graphite and (**b**) nickel manganese cobalt (NMC). The entropic coefficients are also plotted as a function of intercalated *Li*. Here, plotted open circuit potentials are at 25 °C. These OCP profiles, combined with entropic coefficients, define OCP at any temperature

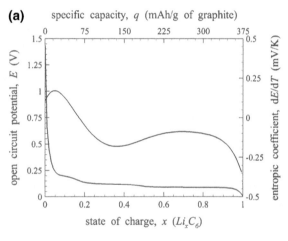

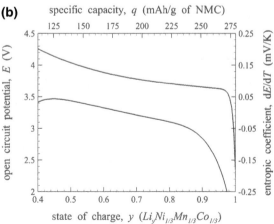

$\Delta G = \Delta H - T\Delta S$. Thus, out of ΔH energy change associated with the reaction, ΔG manifests as available work (or cell potential in the context of electrochemistry) and $T\Delta S$ dissipates as heat. Upon rearrangements, it can be shown that:

$$\Delta S = q \frac{\partial E}{\partial T} \tag{4.4}$$

where $\Delta G = -nFE = -qE$ is the relation between cell potential and free energy. The generated reversible/entropic heat is:

$$Q_{rev} = qT\Delta S = qT\frac{\partial E}{\partial T} \tag{4.5}$$

$$\dot{Q}_{rev} = -JT\frac{\partial E}{\partial T} \tag{4.6}$$

where the expression (4.4) quantifies heat generated when q amount of charge is drawn from the cell, while (4.4) describes rate of heat generation when discharge current is J. By convention, discharge current is considered positive, while charge quantities are taken as negative. Functional relations for OCP and $\partial E/\partial T$ for these materials are tabulated in Table 4.1.

4.2.3 Cell Capacity and C-Rate

The maximum lithium concentration in these materials, material density, and specific capacity are interrelated by the following expression (4.6).

Table 4.1 Open circuit potential and entropic coefficient for graphite and NMC

Property	Functional form
$E(x)$ for graphite [27]	$E(x) = 0.1493 + 0.8493e^{-61.79x} + 0.3824e^{-665.8x}$
	$\quad -e^{(39.24x-41.92)} - 0.03131\tan^{-1}(25.59x - 4.099)$
	$\quad\quad -0.009434\tan^{-1}(32.49x - 15.74)$
$\frac{dE(x)}{dT}$ for graphite [28]	$\frac{dE(x)}{dT} = -58.294x^6 + 189.93x^5 - 240.4x^4 + 144.32x^3$
	$\quad\quad -38.87x^2 + 2.8642x + 0.1079$
$E(y)$ for NMC [28]	$E(y) = -2.5947y^3 + 7.1062y^2 - 6.9922y + 6.0826$
	$\quad\quad -5.4549 \times 10^{-5}e^{(124.23y-114.2593)}$
$\frac{dE(y)}{dT}$ for NMC [28]	$\frac{dE(y)}{dT} = -190.34y^6 + 733.46y^5 - 1172.6y^4 + 995.88y^3$
	$\quad\quad -474.04y^2 + 119.72y - 12.457$

$$q(\text{mAh/g}) = \frac{F(\text{C/mol}) \cdot C^{max}(\text{mol/m}^3)}{\rho(\text{kg/m}^3)} \cdot \frac{1000^{\text{mA/A}}}{3600^{\text{s/hr}}} \cdot \frac{1}{1000^{\text{g/kg}}}$$

$$\therefore q(\text{mAh/g}) = \frac{F(\text{C/mol}) \cdot C^{max}(\text{mol/m}^3)}{3600 \cdot \rho(\text{kg/m}^3)} \tag{4.7}$$

The corresponding values for graphite and NMC are tabulated in Table 4.2. Here, available capacity of NMC is reported as 60% of maximum capacity. This value depends on the voltage window of operation. For the present work, NMC is charged till 4.2 V, which is around 40% lithium content, thus reducing the maximum available capacity to 60%. This is also apparent in Fig. 4.3b.

Specific capacity specifies the upper limit for material utilization. The operation current is also determined from specific capacity. The current for galvanostatic battery operation is specified as C-rate. C-rate is inversely related to the ideal time for battery operation in hours. For example, a C/10 operation implies that ideally, it will take 10 h for the cell to discharge (and similarly charge). The Li-ion cells are usually made cathode limited, i.e., cathode active material loading (in terms of equivalent charge) is smaller than the anode amount. Higher anode amount provides an additional supply of lithium as it gets depleted in SEI growth and also helps to circumvent unwanted phenomena at anode such as Li-plating. Thus, both cell capacity and operating current are determined based on cathode content. For the cell constructed based on material specifications provided in Table 4.1, the current density for 1C operation is:

$$J_{1C} = 167.58 \text{ mAh/g of NMC} \tag{4.8}$$

If the cathode thickness is L_{cat} and the cathode active material volume fraction is ε_{cat}^s, then current density per unit cell cross-sectional area is:

$$J_{1C} = q \cdot \rho_S \cdot \varepsilon_{cat}^s \cdot L_{cat} \text{A/m}^2 \tag{4.9}$$

and corresponding applied current density at a prescribed C-rate is:

$$J_{app} = C - rate \times J_{1C} \tag{4.10}$$

Table 4.2 Lithium storage capacities of graphite and NMC materials

Property	Graphite (Li_xC_6)	NMC $(Li_yNi_{1/3}Mn_{1/3}Co_{1/3})$
Maximum Li concentration (mol/m^3)	30,555	49,500
Material density (kg/m^3)	2200	4750
Specific capacity (mAh/g)	372.25	279.30
Available specific capacity (mAh/g)	$372.25 (x = 0 - 1)$	$167.58 (y = 0.4 - 1)$

4.2.4 Intercalation Kinetics

As mentioned earlier, Li-ion battery is an intercalation chemistry. IUPAC gold book defines intercalation reaction as:

> Reaction, generally reversible, that involves the penetration of a host material by guest species without causing a major structural modification of the host.

In Li-ion battery context, it is modeled as electrochemical reaction at the material surface and solid-state diffusion of Li inside the bulk. Although there have been arguments against the model validity, especially for pulse operation [22–25], this simple model sufficiently predicts the cell behavior for normal operations and is followed in the current work. The dependence of current density at the active material surface on overpotential is assumed to be dictated by the Butler–Volmer relation Eq. (4.11).

$$j = k\left(C_s^f C_e \left(C_s^{max} - C_s^f\right)\right)^{1/2} \left\{e^{\frac{F}{2RT}\eta} - e^{-\frac{F}{2RT}\eta}\right\} \tag{4.11}$$

where $\eta = \phi_s - \phi_e - E\left(C_s^f\right)$ is overpotential, C_s^f is surface concentration of Li in active material, C_e is local electrolyte phase concentration, C_s^{max} is maximum concentration of Li in active material (e.g., Table 4.2), and $E(x) = E\left(C_s^f\right)$ is open circuit potential of active material at a given surface concentration (e.g., Table 4.1).

The solid-state diffusion coefficient is usually several orders smaller than that for electrolyte phase (described shortly hereafter). It can also vary with local lithium concentration, but a constant value is usually adopted for most simulation studies. Typical values are given in Table 4.3.

4.2.5 Electrolyte Transport Properties

The electrolyte is prepared by dissolving suitable Li salt in organic electrolyte. The salt dissolves and produces Li^+ and A^- ions. Correspondingly, if $LiPF_6$ is the salt, it will produce Li^+ and PF_6^- ions, which will be the charge carriers in the electrolyte phase. These ions move around due to their concentration gradients as well as electrolyte phase potential. Thus, charge transport is due to two effects—migration and ionic diffusion. These transport processes are modeled by the following flux relations:

Table 4.3 Solid-state diffusion coefficient for graphite and NMC [27]

Material	Solid-state diffusivity (m²/s)
Graphite	1.6×10^{-14}
NMC	3.0×10^{-14}

$$J_D = -D\nabla C \tag{4.12}$$

$$J_\kappa = -\kappa\nabla\phi_e \tag{4.13}$$

$$J_{\kappa_D} = -\kappa_D\nabla\ln C \tag{4.14}$$

where J_D, J_κ, and J_{κ_D} are species diffusive flux, current flow due to migration, and current flow due to ionic diffusion, respectively. D is Fickian diffusivity, κ is ionic conductivity, and κ_D is diffusional conductivity. Diffusional conductivity is related to ionic conductivity via the following expression:

$$\kappa_D = \frac{2RT\kappa}{F}(t^+ - 1)\left(1 + \frac{d\ln f_\pm}{d\ln C}\right) \tag{4.15}$$

Here, t^+ is referred to as transference number. As Li-ion battery electrolytes employ larger salt concentrations, use of concentrated solution theory becomes necessary to describe the electrolyte phase transport. In particular, expression (4.15) and concentration dependence of transport properties (e.g., Table 4.4) describe such a behavior.

As observed from Figs. 4.4 and 4.5, the electrolyte phase transport properties are a strong function of both salt concentration and temperature. Valøen and Reimers [29] report measurements for these transport properties for Li-ion battery electrolyte made up of $LiPF_6$ in $PC/EC/DMC$ solvent (propylene carbonate/ethylene carbonate/dimethyl carbonate). These plots of transport property variations reveal that diffusivity reduces with Li^+ concentration increase at a constant temperature. This stems from concentrated solution effects. The dilute solution diffusivity represents the hindrance offered by solvent molecules to ionic movement. As concentration of the solute increases, the solute molecules offer additional hindrance, leading to reduced ionic mobility, which is quantified in terms of reduced diffusivity. A non-monotonic trend is obtained for conductivity with the maxima around

Table 4.4 Functional dependence of Li^+ diffusivity, ionic conductivity, and diffusional conductivity measured for $LiPF_6$ in $PC/EC/DMC$ electrolyte as a function of Li^+ concentration and temperature [29]

Diffusivity, $D(\text{m}^2/\text{s})$	$\log_{10} D = -8.43 - \left(\frac{54}{T-(229+5(C/1000))}\right) - 0.22(C/1000)$
Ionic conductivity, $\kappa(\text{s/m})$	$100\sqrt{\dfrac{\kappa}{C}} = -10.5 + 0.0740T - 6.96\times10^{-5}T^2$ $+ 0.668(C/1000) - 0.0178(C/1000)T$ $+ 2.80\times10^{-5}(C/1000)T^2 + 0.494(C/1000)^2$ $- 8.86\times10^{-4}(C/1000)^2 T$
Diffusional conductivity, κ_D (A/m)	$\dfrac{\kappa_D}{\kappa} = -\dfrac{2RT}{F}\left\{\begin{array}{c} 0.601 - 0.24(C/1000)^{1/2} \\ + 0.982(1 - 0.0052(T - 294))(C/1000)^{3/2} \end{array}\right\}$

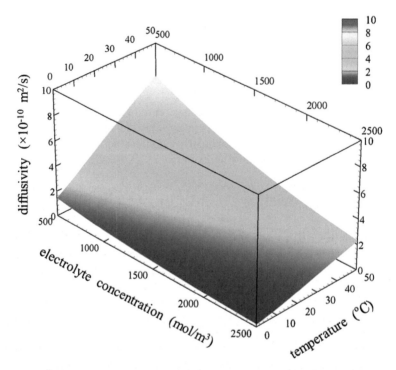

Fig. 4.4 Diffusivity for Li^+ in electrolyte made up of $LiPF_6$ in $PC/EC/DMC$ as a function of ionic concentration as well as temperature

1000 mol/m^3 ionic concentration. Consequently, the electrolyte is typically prepared with around 1 molar salt concentration. The temperature trend is rather simple, offering better transport at elevated temperatures. This can be understood from increased kinetic energies of ions (kT) leading to more agile ions in the solution.

4.3 Modeling Isothermal Operation of a Li-Ion Cell

The operation of a LIB involves multiple transport processes simultaneously taking place in different parts of the cell. The electrons flow in external circuit, current collector, and solid phase of electrodes. At electrode–electrolyte interface, lithium intercalation–deintercalation takes place, leading to availability of Li atoms and e^- in the solid phase, and Li^+ in the electrolyte phase. Li^+ ions travel from one electrode to the another via electrolyte. The macrohomogeneous model (that is built up on the porous electrode theory) provides an inclusive description of this physical picture.

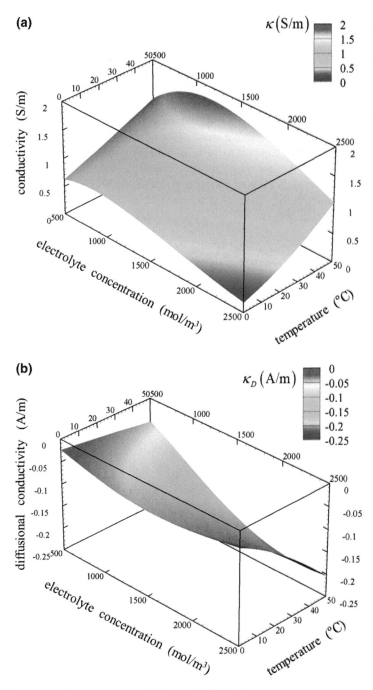

Fig. 4.5 Conductivities reported for $LiPF_6$ in $PC/EC/DMC$ electrolyte as a function of Li^+ concentration and temperature. **a** Ionic conductivity κ and **b** diffusional conductivity κ_D

4.3.1 Macrohomogeneous Description

The porous nature of components constituting a typical LIB necessitates appropriate transformation of clear media transport laws to their porous media counterparts by utilization of volume-averaging techniques. Newman and coworkers [9–11] were the first to develop the corresponding theory in the context of electrochemical systems, often referred to as the porous electrode theory.

Intercalation in the solid active material is modeled as a combined effect of diffusion inside the bulk and electrochemical reaction at the surface. The electrodes are composed of active material particles. These particles are approximated as spheres, and the concentration evolution is governed by Fick's second law:

$$\frac{\partial C_s}{\partial t} = \frac{D_s}{r^2} \frac{\partial}{\partial r} \left(r^2 \frac{\partial C_s}{\partial r} \right) \tag{4.16}$$

along with appropriate boundary conditions:

$$\text{At } r = 0, \frac{\partial C_s}{\partial r} = 0 \tag{4.17}$$

$$\text{At } r = R_s, -D\frac{\partial C_s}{\partial r} = j_s/F \tag{4.18}$$

where j_s is current flux per unit surface area of the particle (also known as surface current density).

Using concentrated solution theory for binary electrolytes and porous media approximation, the evolution of Li^+ concentration in electrolyte phase is modeled via the following expression:

$$\varepsilon \frac{\partial C_e}{\partial t} = \frac{\partial}{\partial x} \left(D \frac{\varepsilon}{\tau} \frac{\partial C_e}{\partial x} \right) + \left(\frac{1 - t^+}{F} \right) J \tag{4.19}$$

Here, ε is porosity, τ is tortuosity, $D = D(C_e)$ is diffusivity, t^+ is transference number, and J is volumetric current source term due to intercalation reaction. Note that local charge neutrality dictates that the concentration of Li^+ and corresponding anion is stoichiometrically equal at all times. The ionic concentration can change due to diffusion (the first term), migration (in response to electric field established in electrolyte phase $-t^+ J/F$), and generation due to volumetric reaction (J/F).

Similarly, the statement of conservation of charge in the electrolyte phase and solid phase can be, respectively, written as follows:

$$\frac{\partial}{\partial x} \left(\kappa \frac{\varepsilon}{\tau} \frac{\partial \phi_e}{\partial x} \right) + \frac{\partial}{\partial x} \left(\kappa_D \frac{\varepsilon}{\tau} \frac{\partial \ln C_e}{\partial x} \right) + J = 0 \tag{4.20}$$

$$\sigma^{eff} \frac{\partial^2 \phi_s}{\partial x^2} = J \tag{4.21}$$

where σ^{eff} is effective electronic conductivity of solid phase.

Here, the volumetric current source term J is related to particle surface flux. The surface current density j_s represents current density per unit surface of the active material particle. For a m^2 of active material surface area per unit electrode volume, the two are correlated via specific surface area a as:

$$J = aj_s \tag{4.22}$$

Equations (4.16), (4.19), (4.20), and (4.21) represent formulation of different transport mechanisms and together define the LIB dynamics. The boundary conditions are dependent on the operating mode of the LIB. For constant-current (CC) operation, boundary condition is specified as fixed charge flux, while, for constant-voltage (CV) operation, the boundary condition is in terms of constant-voltage difference.

4.3.2 Comments on Mathematical Nature of Governing Equations and Solution

It should be noted that the system of Eqs. (4.16), (4.19), (4.20), and (4.21) contains two different spatial dimensions. The particle behavior is a function of radial dimension, while the remaining equations demonstrate dependence along the thickness direction. Also, the particle behavior must be quantified for each of the particles along the thickness direction. All the equations are coupled via volumetric current source term J and particle surface flux j_s, which in turn depend on particle surface concentration as well as local electrolyte concentration. Each of these variables is solved on a different subdomain. The situation is schematically presented in Fig. 4.6.

The solution of above set of equations presents an interesting computational complexity. Both the concentration Eqs. (4.16) and (4.19) are second-order parabolic partial differential equations and are accompanied by suitable initial and boundary conditions. However, the potential expressions (4.20) and (4.21) are elliptic partial differential equations with nonlinear source terms (source term exhibiting nonlinear dependence on primary-dependent variable). This necessitates the solution of these equations at each time step. The problem arises from the fact that all the physically realistic boundary conditions for these equations are Neumann type and a Poisson equation (or a set of Poisson equations) with all

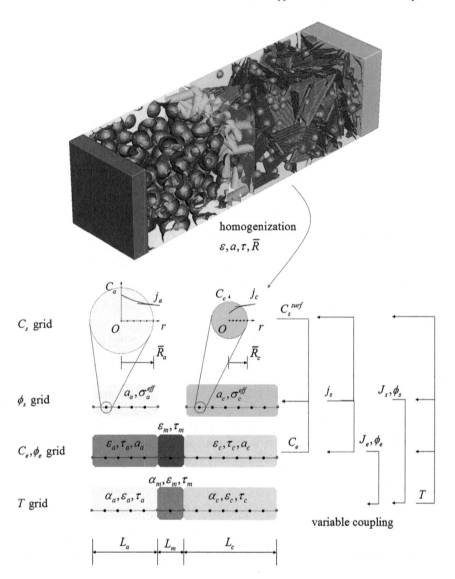

Fig. 4.6 Dynamic behavior of porous components forming a lithium ion battery can be simplified via homogenization in the spatial coordinates (porous electrode theory). The equivalent behavior is represented by solid-state diffusion in active material particles, and electrolyte phase concentration and potential along the electrode thickness. Each of these subdomains (represented by different colors) is identified by their corresponding porous media properties, i.e., active area a, effective electronic conductivity σ^{eff}, porosity ε, and tortuosity τ. Various physical variables are solved on different subdomains and are all tightly coupled via reaction source terms. Red dots represent boundary and/or interface conditions

Table 4.5 Physics-based boundary conditions for nonlinear Poisson equations quantifying charge conservation [Eqs. (4.20) and (4.21)]

Variable	Anode–current collector boundary $(x = 0)$	Anode–separator boundary $(x = L_a)$	Cathode–separator boundary $(x = L_a + L_m)$	Cathode–current collector boundary $(x = L_a + L_m + L_c)$
Anode potential, ϕ_a	$J_{app} = -\sigma_a^{eff} \frac{\partial \phi_a}{\partial x}$	$\frac{\partial \phi_a}{\partial x} = 0$	–	–
Cathode potential, ϕ_c	–	–	$\frac{\partial \phi_c}{\partial x} = 0$	$J_{app} = -\sigma_c^{eff} \frac{\partial \phi_c}{\partial x}$
Electrolyte potential, ϕ_e	$\frac{\partial \phi_e}{\partial x} = 0$	Flux continuity	Flux continuity	$\frac{\partial \phi_e}{\partial x} = 0$

Neumann conditions represents an ill-posed problem. The corresponding physical boundary conditions are listed in Table 4.5.

To solve this system of equations, one Dirichlet boundary condition has to be specified. Electrolyte phase potential is arbitrarily set to zero at either anode–current collector boundary (or cathode–current collector boundary). This makes the system of equations constitute a well-posed problem. Moreover, changing this boundary condition does not alter the physics of the system, since the equations are coupled via volumetric current source term J. The conservation property of elliptic equations ensures that the summation of all the volumetric source terms is equal to the applied current density. This also assures that the amount of charge is balanced in all the different phases (both electrodes and electrolyte). When the equations are solved, the boundary which is assigned $\phi_e = 0$ also displays $\partial \phi_e / \partial x = 0$ due to charge conservation (refer to Sect. 4.3.3).

4.3.3 Results and Discussion

As discussed earlier, the theoretical bounds on cell performance are determined from thermodynamic behavior of electrode materials. Once the anode and cathode active material contents are fixed, one can relate state of charges x and y for individual electrodes with cell capacity q. Using this relation, one can define thermodynamic open circuit cell voltage from corresponding open circuit potentials of electrode materials. The dashed line in Fig. 4.7a represents OCV of the cell studied.

$$E_{cell}(q) = E_{cathode}(y) - E_{anode}(x) \tag{4.23}$$

Based on active material content in electrodes, 1C current density is defined using expression (4.8). This section reports results of a graphite–NMC Li-ion cell having 40 μm electrode thicknesses and 20 μm separator thickness. Theoretical cell capacity comes out to be 19.5 Ah/m^2. Figure 4.7a reports the terminal cell voltage

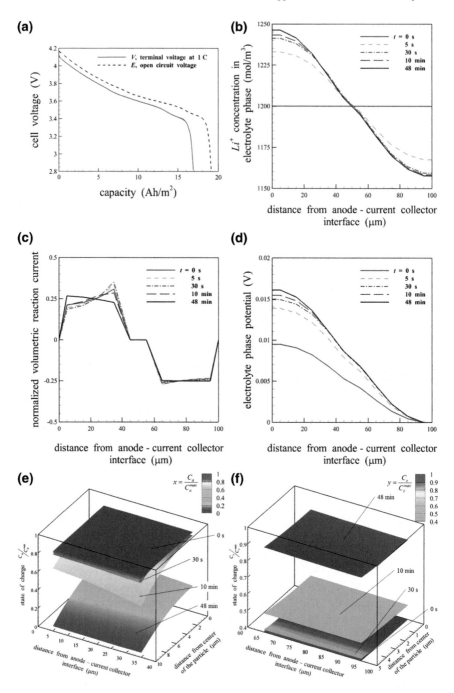

◀**Fig. 4.7** Isothermal constant-current discharge behavior of a Li-ion cell composed of graphite anode and nickel manganese cobalt oxide cathode. (**a**) Terminal voltage versus discharge capacity at 1C current. Evolution of (**b**) Li^+ concentration in electrolyte. (**c**) Normalized reaction current density. (**d**) Electrolyte phase potential as a function of distance along cell thickness, and normalized Li concentration (**e**) in anode and (**f**) in cathode as a function of radial and axial direction at different times. Due to transport limitations, the cell discharge ends at 48 min (theoretical discharge time is 60 min for 1C discharge)

as a function of discharge capacity at 1C. There are two observable details: The cell terminal voltage V is always smaller than open circuit cell voltage E; the discharge capacity is smaller than theoretical cell capacity. The difference is due to finite rate transport limitations.

At time $t = 0$, there exists no concentration gradients. This is displayed in Fig. 4.7b, e, f. The normalized reaction current density and electrolyte phase potential are not zero on $t = 0$ plots, since the solution algorithm solves potential equations with applied current density boundary condition. The normalized reaction current density j^* is defined such that its summation over electrode thickness is unity, i.e.

$$j^* = \frac{J \Delta x}{J_{app}} \qquad (4.24)$$

During discharge, reaction current density is positive at anode and negative in cathode.

As time is increased, concentration gradients establish in both electrolyte phase and solid phase and reach their maximum gradient situation toward the end of discharge. This local change in concentration profile also changes the current density distribution, but always such that the integration over respective electrode thicknesses is unity.

The discharge stops when either anode particles are nearly empty or cathode particles are nearly full, as dictated by particle surface concentrations. At finite rates of operation, the radial concentration gradients in the particle are maximum toward the end of operation (as revealed by Fig. 4.7e, f). The gradients are steeper for larger particles. In other words, particle utilization is better for a smaller size active material particle.

Figure 4.8 presents cell behavior at different rates of operation. As discharge rate is decreased, cell behavior approaches its thermodynamic limit. Alternatively, as operating current density is increased, active particle utilization drops due to steeper concentration gradients. This leads to reduced capacity. Moreover, to maintain higher rates of discharge, kinetic overpotential increases, leading to greater overpotential and correspondingly smaller cell terminal voltage. The concentration profiles in electrolyte phase and active material particles are plotted at the end of discharge in Fig. 4.8b, e, f. The profiles exhibit increased gradients at higher rates of operation. Similar behavior is also observed for electrolyte phase potential (Fig. 4.8d). As expected, Fig. 4.8c demonstrates more uniform reaction current density at lower rates of operation.

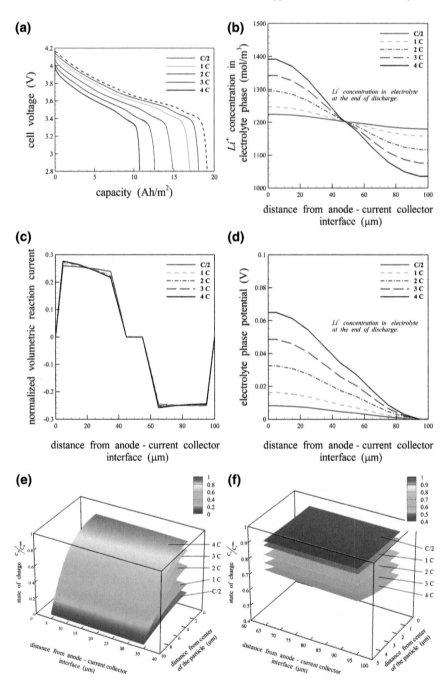

Fig. 4.8 Isothermal discharge of graphite–NMC lithium ion cell at different discharge current densities. **a** Evolution of cell terminal voltage as a function of discharge capacity. **b** Li^+ concentration in electrolyte phase, **c** normalized reaction current density. **d** Electrolyte phase potential along $x-$ direction at final discharge time. **e** and **f** depict dimensionless Li concentration along radial and axial direction to final discharge times, in anode and cathode, respectively

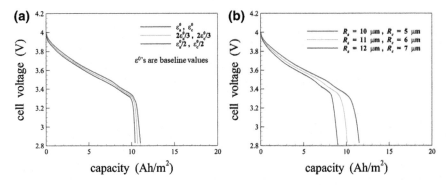

Fig. 4.9 Cell terminal voltage as a function of discharge capacity at 1C constant-current operation for (**a**) different electrode porosities and (**b**) different active material particle radii. Smaller porosities lead to greater concentration gradients in the electrolyte phase, which leads to reduced cell capacity. Larger particle size gives rise to increased concentration overpotential and equivalently reduced discharge capacity

If electrode porosity is reduced, it increases the resistance offered to electrolyte phase transport and which in turn leads to reduced capacity. Figure 4.9a presents results of 1C discharge of this cell for different electrode porosities.

If active material particle size is increased, it leads to greater concentration overpotential in solid phase and higher particle underutilization. This gives rise to both smaller cell terminal voltage and decreased discharge capacity.

If electrode thickness is increased, it leads to increased active material content. This in turn leads to more capacity for same C-rate operation (note that changing electrode thickness changes 1C current density). Figure 4.10 reports effect of electrode thickness on cell operation at 1C. Thicker electrodes also lead to increased concentration gradients along thickness directions, which also gives rise to more inefficient utilization of the electrode material. This is apparent from comparing normalized cell capacities, as reported in Table 4.6. This deterioration in electrode

Fig. 4.10 Effect of electrode thickness on performance of a graphite–NMC Li-ion cell. The cells are discharged at 1C current for different electrode thicknesses **a** 40 μm, **b** 60 μm, **c** 80 μm. Larger thickness electrodes have more active material and equivalently greater capacity. Larger electrodes also exhibit greater overpotentials, which leads to lesser electrode utilization at the same C-rate

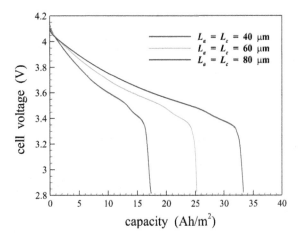

Table 4.6 Normalized cell capacities for 1C discharge operation of cells with different electrode thicknesses

Electrode thickness (μm)	Theoretical cell capacity (Ah/m^2)	Discharge capacity at 1C (Ah/m^2)	Normalized cell capacity (%)
40	19.50	16.10	82.56
60	29.25	24.06	82.26
80	39.00	31.79	81.51

utilization for thicker electrodes is more apparent at higher operational current densities.

4.4 Importance of Thermal Effects and Its Influence on Electrochemical Operation

Due to thermodynamic limitation on heat-to-work conversion, only a part (ΔG) of reaction heat (ΔH) is available as work. The remaining part is dissipated as heat. This entropic component ($T\Delta S$) can lead to temperature changes during cell operation.

At equilibrium, no current flows and the various potentials (solid phase potential, electrolyte phase potential, and open circuit potential) are in balance. To draw a finite amount of current, an imbalance has to be created in terms of these potentials. Overpotential η is a measure of this imbalance. Alternatively, overpotential can be understood as the penalty to be paid to proceed the electrochemical reactions in either direction. This energy penalty $j\eta$ manifests as heat generation at the electrode–electrolyte interface. Sign of overpotential is always the same as sign of current density [Expression (4.10)]; hence, this term always amounts to heat addition.

Moreover, when current flows across potential difference, it leads to Joule heating. For electronic current flow, volumetric heat generation is given as $\sigma_s^{eff} \nabla \phi_s \cdot \nabla \phi_s$. Ionic current is made up of two components—migration and diffusion—and correspondingly has two different heat source terms $\kappa^{eff} \nabla \phi_e \cdot \nabla \phi_e$ and $\kappa_D^{eff} \nabla \ln C_e \cdot \nabla \phi_e$.

The first term is reversible in nature and can assume positive or negative sign depending on the direction of current flow. The remaining terms are purely dissipative leading to heating of the cell. Depending on the ambient conditions and cell exposure, a part of generated heat is carried away.

If the cell dimensions are small enough, e.g., coin cell, then the heat generation due to electrochemical operation is not significant to raise cell temperature. For medium (e.g., cylindrical cell) or larger (e.g., pouch cell, battery pack) size cells, the heat generation is high enough to appreciably change cell temperature [30]. The present section outlines the modeling strategy to incorporate thermal effects with electrochemical operation of a Li-ion cell.

4.4.1 Energy Equation to Define Temperature Changes

First law of thermodynamics quantifies the energy changes associated with heat and work interactions for any system. For an electrochemical system (such as a Li-ion battery), heat generation due to electrochemical operation and heat dissipation due to convection and electrochemical work are major system–surroundings interactions.

At the pore scale, material heterogeneity is observed inside the porous electrodes and each of these materials has different thermal properties. The thermal diffusivities [31] range from $\mathcal{O}(10^{-7})$ for graphite and $LiCoO_2$ to $\mathcal{O}(10^{-4})$ for copper and aluminum (SI units—m^2/s). This may suggest the existence of temperature difference across material interfaces at the pore scale. But in comparison with other transport processes, e.g., solid-state diffusion ($D \sim \mathcal{O}(10^{-14})$ m^2/s) and electrolyte phase transport ($D \sim \mathcal{O}(10^{-10})$ m^2/s), thermal transport is orders of magnitude faster, justifying the assumption of local thermal equilibrium. Thus, one can define a unique temperature field of the porous electrode, rather than treating each of the phases individually. This still does not exclude the possibility of temperature gradients across porous electrodes in a volume-averaged sense. This simplification leads to the energy equation for a control volume in a Li-ion battery:

$$(\rho C_P)_e \frac{\partial T}{\partial t} = \nabla \cdot (k_e \nabla T) + \dot{Q} \tag{4.25}$$

where subscript 'e' denotes equivalent properties—defined for porous composite involving solid phases as well as electrolyte. Note that these equivalent properties have to be computed for a given electrode composition, porosity, and choice of electrolyte. Depending on the location of the control volume, the volumetric heat generation term adopts different forms [27]:

In electrodes:

$$\dot{Q} = -jT \frac{\partial E}{\partial T} + j\eta + \sigma_s^{eff} \nabla \phi_s \cdot \nabla \phi_s + \kappa^{eff} \nabla \phi_e \cdot \nabla \phi_e$$
$$+ \kappa_D^{eff} \nabla \ln C_e \cdot \nabla \phi_e \tag{4.26}$$

In separator:

$$\dot{Q} = \kappa^{eff} \nabla \phi_e \cdot \nabla \phi_e + \kappa_D^{eff} \nabla \ln C_e \cdot \nabla \phi_e \tag{4.27}$$

In current collector:

$$\dot{Q} = \sigma_s^{eff} \nabla \phi_s \cdot \nabla \phi_S \tag{4.28}$$

The heat exchange with ambient occurs via boundary condition:

$$-k_e \nabla T \cdot \hat{n} = h(T - T_\infty) \tag{4.29}$$

For large size cells (or modules), energy equation has to be solved in this distributed fashion. For a medium size cell (e.g., an 18,650 cylindrical cell), usually the Biot number $\text{Bi} = h\mathcal{L}/k_e \ll 1$, which quantifies the surface convection against bulk conduction effects. Consequently, the heat conduction inside the cell becomes insignificant since the material has high thermal conductivity leading to negligible temperature gradients inside the cell. In other words, the entire cell can be treated with one representative temperature value that evolves in time. In such an event, the energy equation can be integrated over spatial coordinate, and with the help of boundary condition (4.28), global energy balance becomes:

$$mC_p \frac{dT}{dt} = \dot{Q} - hA_s(T - T_\infty) \tag{4.30}$$

along with integrated source term:

$$
\begin{aligned}
\frac{\dot{Q}}{A_{\text{electrode}}} = {} & \int_0^{L_a} \left(-jT \frac{\partial E_a}{\partial T} + j\eta + \sigma_a^{eff} \nabla \phi_a \cdot \nabla \phi_a \right) dx \\
& + \int_0^{L_a} \left(\kappa^{eff} \nabla \phi_e \cdot \nabla \phi_e + \kappa_D^{eff} \nabla \ln C_e \cdot \nabla \phi_e \right) dx \\
& + \int_{L_a}^{L_a + L_m} \left(\kappa^{eff} \nabla \phi_e \cdot \nabla \phi_e + \kappa_D^{eff} \nabla \ln C_e \cdot \nabla \phi_e \right) dx \\
& + \int_{L_a + L_m}^{L_a + L_m + L_c} \left(-jT \frac{\partial E_c}{\partial T} + j\eta + \sigma_c^{eff} \nabla \phi_c \cdot \nabla \phi_c \right) dx \\
& + \int_{L_a + L_m}^{L_a + L_m + L_c} \left(\kappa^{eff} \nabla \phi_e \cdot \nabla \phi_e + \kappa_D^{eff} \nabla \ln C_e \cdot \nabla \phi_e \right) dx
\end{aligned}
\tag{4.31}
$$

Depending on the resistance offered by other inactive components (e.g., current collector) to the current flow, additional $I^2 R$ terms should have to be included in the expression (4.31).

This lumped approximation of the cell thermal behavior implies that the cell temperature rises uniformly across the cell. If dimensions of the cell are large enough to violate the small Biot number approximation, a detailed multidimensional analysis [utilizing Eq. (4.25)] for cell temperature is required.

The rise of cell temperature affects the transport coefficients. The temperature dependence of electrolyte properties is reported earlier (Fig. 4.4 and 4.5; Table 4.4). The solid-state diffusivity and reaction rate constants exhibit an

Arrhenius type dependence on temperature. Each of these properties increase with increase in temperature, thus facilitating better transport. The coupling with energy equation and modification in solution strategy are schematically presented in Fig. 4.6. This heat generation due to electrochemical transport can be sufficient enough to appreciably raise cell temperature to allow for cold-start of Li-ion cells [27]. On the other hand, the temperature cannot be allowed to be very high since that might trigger thermal runaway [32–34] or cause other deleterious effects.

4.4.2 Results and Discussion

Thermal response of an 18,650 cylindrical Li-ion cell with graphite anode and NMC cathode is studied in the present section. The electrodes are both 80 μm thick. As explained earlier, a lumped thermal model can be used to simulate thermal response of such cells. The temperature evolution in such cells falls between two extremes—adiabatic and isothermal. In an adiabatic cell, the heat transfer coefficient is 0 and the total heat generated due to electrochemical reaction goes into increasing cell temperature. On the other hand, if heat transfer coefficient is very large, the generated heat gets dissipated via convection for the most part, and the cell temperature stays fairly constant. For most practical cases, the cell response falls between these two extremes.

Figure 4.11a shows the cell performance and temperature rise as a function of discharge rate for adiabatic cells (i.e., maximum temperature rise at the end of operation). Terminal voltage decreases with C-rate due to greater overpotential at higher rates.

Figure 4.11b demonstrates the cell performance at 2C discharge rate for different convective cooling conditions. As heat transfer coefficient increases, the temperature rise is lesser and cell capacity is also smaller, as higher temperature leads to better transport coefficients and higher capacities.

Figure 4.11c shows the effect of changing electrode porosities on temperature rise and discharge performance. Decreasing electrode porosity leads to higher internal resistance, which in turn gives more heat generation and higher temperature rise. Reduced porosity also gives rise to greater concentration gradients and reduced cell capacity. Capacity increase due to increased temperature and capacity decrease due to reduced porosity can compensate each other, as reflected from the lower porosity simulations on plot (c).

4.5 Direct Numerical Simulation

The porous electrode theory-based description of electrochemical operation of lithium batteries stands on many assumptions such as spherical particle geometry and conventional correlations among porosity, tortuosity, active area, etc. Along

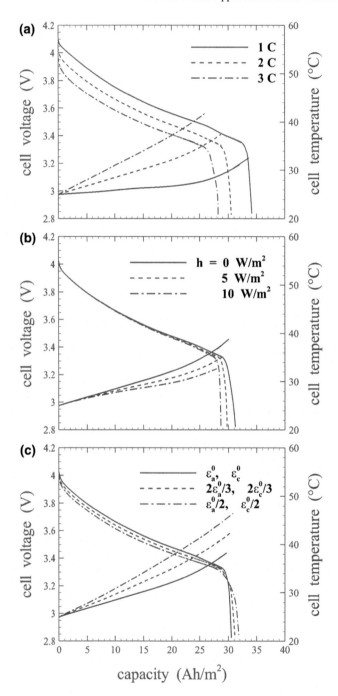

capacity (Ah/m²)

◀**Fig. 4.11** Discharge performance of a graphite–NMC Li-ion cell along with temperature rise due to heat generated from cell internal resistance. Cell terminal voltage and temperature rise for (a) adiabatic discharge at different current densities, (b) 2C cell discharge with different ambient convection coefficients, (c) 2C adiabatic discharge for cells with different electrode porosities and hence different internal resistances. **a** As operating current is increased, irreversible (or Joule) $I^2 R_{in}$ heating increases, which leads to increased cell temperature and enhanced transport. **b** As convection coefficient is increased, cell temperature rise is smaller at the same rate of operation. In the extreme event ($h \rightarrow \infty$), cell temperature is constant (at ambient) and leads to smallest capacity compared to adiabatic situation ($h = 0$), where highest temperature rise gives rise to enhanced transport and best performance for a given set of electrochemical operation parameters. **c** Smaller porosities lead to greater internal resistance and correspondingly higher irreversible heat generation

with the progress in battery design, invention of new materials, and modification in electrode manufacturing strategies, many of these assumptions are under scrutiny.

With the present electrode thicknesses and particle dimensions, representative elementary volume dimensions cannot always be met. The electrode thickness is still large enough to be simulated with porous electrode theory but when discretization is performed along the thickness of the electrodes, each of the discretized volumes is not sufficiently large to apply porous approximation.

4.5.1   Governing Equations

The most straightforward answer to resolving all these issues is to let go of the porous electrode theory and solve for each of the variables at pore scale. This would modify the set of governing equations as follows [35]:

Solid-state diffusion:

$$\frac{\partial C_s}{\partial t} = D \nabla^2 C_s \tag{4.32}$$

Li^+ *transport in electrolyte phase*:

$$\frac{\partial C_e}{\partial t} = \nabla \cdot (D \nabla C_e) + \frac{t^+}{F} \nabla . J_e \tag{4.33}$$

Electrolyte phase current:

$$J_e = -\kappa \nabla \phi_e - \kappa_D \nabla \ln C_e \tag{4.34}$$

Charge conservation in electrolyte phase (ionic current):

$$\nabla \cdot (\kappa \nabla \phi_e + \kappa_D \nabla \ln C_e) = 0 \tag{4.35}$$

Charge conservation in solid phase (electronic current):

$$\nabla \cdot (\sigma \nabla \phi_s) = 0 \tag{4.36}$$

In addition to flux continuity, the governing equations are coupled through electrode–electrolyte interface via following relations:

$$-D\nabla C_s \cdot \hat{n} = -D\nabla C_e \cdot \hat{n} = j/F \tag{4.37}$$

$$-\sigma \nabla \phi_s \cdot \hat{n} = -(\kappa \nabla \phi_e + \kappa_D \nabla \ln C_e) \cdot \hat{n} = j \tag{4.38}$$

where $\hat{n}$ is local surface normal pointing from solid phase toward electrolyte phase, and j is Butler–Volmer current density.

Though being intuitively simpler, this approach enforces a huge amount of computational burden. Moreover, the geometry at the pore scale is quite complicated. More importantly, generation of realistic pore-scale geometry is quite a challenge.

4.5.2 Microstructural Effects: Properties for Composite Electrodes

Most of the shortcomings of the porous electrode theory-based description are related to accurate quantification of microstructural porous media properties. For example, electrolyte phase tortuosity is usually correlated with porosity via Bruggeman relation:

$$\tau = 1/\sqrt{\varepsilon} \tag{4.39}$$

Electrochemically active area is related to particle radius and volume fraction as:

$$a = \frac{3\varepsilon_s}{R_s} \tag{4.40}$$

Bruggeman relation is based on the assumption of randomly distributed spherical particles [37–40]. On the other hand, the expression for electrochemically active area assumes that the active material particles are spherical and disconnected and entire surface is available for intercalation. Both of these assumptions usually do not hold. If the electrode properties are quantified based on realistic electrode structures, the governing equations do not have to be modified and the same set of equations can still simulate battery performance.

Figure 4.12 reports simulations based on such an approach [36]. In a typical composite electrode, active material particles have dimensions of the order of 5–10 μm, conductive additives are roughly 50–100 nm in diameter, while electrode is typically 50–100 µm thick (binder is a polymeric phase). This allows one to do successive approximations using two different representative elementary volumes. The first—smaller REV—characterizes the electrical conductivity of binder + conductive additive domain, while the second larger REV models active material particles coated with binder + conductive additive domain. The pore space is apparent at this length scale. The figure shows REVs for different composition of solid phase [active material: (binder + conductive additive)] and same void phase porosity. It is interesting to note that for the same value of porosity (30% for the Fig. 4.12), tortuosity increases with increase in secondary phase content, as is

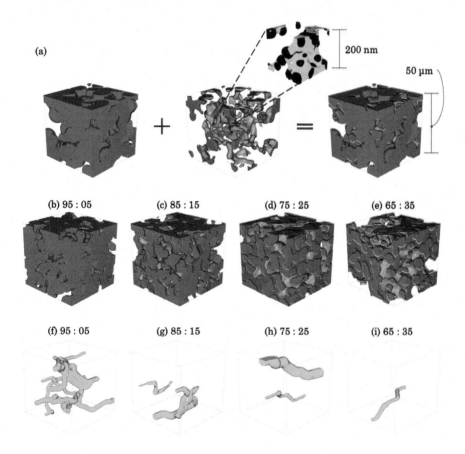

Numbers represent %wt in the order
Active Material : Active Binder (AB + PVDF)

Fig. 4.12 a Composite electrode contains multiple solid phases: active material particles (~5–10 µm), conductive additive (~50–100 nm), and polymeric binder. The electrode dimension is 50–100 µm. This disparity in length scales allows one to perform effective medium approximations at different length scales [36]. Representative electrode microstructures for different composition of active material: (binder + conductive additive) domains (**b**)–(**e**) and corresponding percolation pathways (**f**)–(**i**)

apparent by reduction in number of percolation pathways. Moreover, increasing secondary phase content also decreases the available electrochemically active area, thus leading to limitations due to charge transfer at electrode–electrolyte interface.

The effective properties of such electrode representative elementary volumes are quantified by solving appropriate governing equations at the pore scale, e.g., tortuosity is characterized by solving steady-state concentration equation in void phase, while effective electronic conductivity is obtained by solving potential equation in solid phase along with appropriate bulk conductivities assigned to different material phases.

The changes in amount of different solid phases lead to non-monotonic trends in cell performance. At lower discharge rates (below 1C), limitations arise from kinetic overpotential, while transport resistances become important at higher rates. If enough conductive additives are not present in the porous cathodes, it leads to additional voltage drop due to electron conduction in the solid phase and this drop manifests more at higher discharge rates. For example, a low conductive additive to binder ratio (CA/B = 0.15 by wt.) demonstrates inferior performance compared to a higher ratio (CA/B = 0.85) at 5C, while both of them have very similar performance at lower rates (e.g., C/2). These trends (Fig. 4.13a) change at a smaller active material composition [65% by wt. Fig. 4.13b)] as more amount of conductive additives are present per unit volume of the cathodes (for the same CA/B ratio). Once sufficient percolation is achieved for electronic pathways, and further increase in effective electronic conductivity of the electrodes does not improve performance (Fig. 4.13b). A second observation from comparison of Figs. 4.13a, b is that for lower active material loading, the electrochemically active area reduces which gives rise to increased kinetic overpotential and in turn to reduced voltage, capacity, and energy density. Thus, the electrode composition and porosity decide the effective medium properties (namely, electrochemically active area, tortuosity, and porosity), which in turn influences the cell performance.

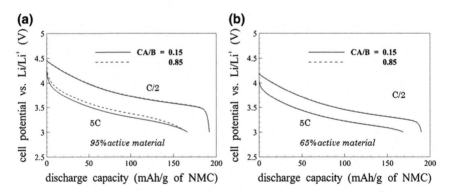

Fig. 4.13 Effect of conductive additive (AB) to binder (PVDF) ratio on cell performance at two different active material loadings **a** 95% by wt. **b** 65% by wt. (electrode porosity is fixed at 30% for these different electrode structures). Lower conductivity composition demonstrates additional transport resistances at higher discharge rates and hampers the rate capability

4.6 Summary and Outlook

The present chapter outlines the modeling attempts in order to comprehend the electrochemical behavior of lithium ion batteries. State-of-the-art LIBs are composed of porous electrodes in order to increase surface area-to-volume ratio and give rise to batteries with better volume-specific and weight-specific energy and power densities.

Porous electrode theory-based description helps explain many dynamic features of cell performance. For small format cells, the electrochemical operation is performed isothermally since heat generation is not large enough to cause appreciable temperature rise, while, for medium to large size batteries, thermal effects become important.

The electrodes for such batteries contain multiple solid phases, and accurate quantification of electrode properties becomes a very important topic. In such an event, direct numerical simulations help connect these novel electrode properties with established porous electrode description.

Acknowledgements The authors of this chapter acknowledge American Society of Mechanical Engineers for the figures reproduced in this chapter from the referenced publications of their journal.

References

1. J.-M. Tarascon, M. Armand, Issues and challenges facing rechargeable lithium batteries. Nature **414**(6861), 359–367 (2001)
2. M.S. Whittingham, Lithium batteries and cathode materials. Chem. Rev. **104**(10), 4271–4302 (2004)
3. M. Armand, J.-M. Tarascon, Building better batteries. Nature **451**(7179), 652–657 (2008)
4. C. Daniel, Materials and processing for lithium-ion batteries. JOM **60**(9), 43–48 (2008)
5. B. Dunn, H. Kamath, J.-M. Tarascon, Electrical energy storage for the grid: a battery of choices. Science **334**(6058), 928–935 (2011)
6. E.J. Cairns, P. Albertus, Batteries for electric and hybrid-electric vehicles. Annu. Rev. Chem. Biomol. Eng. **1**, 299–320 (2010)
7. T.S. Arthur et al., Three-dimensional electrodes and battery architectures. MRS Bull. **36**(07), 523–531 (2011)
8. M. Stein, A. Mistry, P.P. Mukherjee, Mechanistic understanding of the role of evaporation in electrode processing. J. Electrochem. Soc. **164**(7), A1616–A1627 (2017)
9. J. Newman, K.E. Thomas-Alyea, *Electrochemical Systems* (Wiley, Hoboken, 2012)
10. J. Newman, W. Tiedemann, Porous-electrode theory with battery applications. AIChE J. **21**(1), 25–41 (1975)
11. J.S. Newman, C.W. Tobias, Theoretical analysis of current distribution in porous electrodes. J. Electrochem. Soc. **109**(12), 1183–1191 (1962)
12. R. Spotnitz, J. Franklin, Abuse behavior of high-power, lithium-ion cells. J. Power Sources **113**(1), 81–100 (2003)
13. D. Doughty, E.P. Roth, A general discussion of Li ion battery safety. Electrochem. Soc. Interface **21**(2), 37–44 (2012)
14. C. Mikolajczak et al., *Lithium-ion Batteries Hazard and Use Assessment* (Springer Science & Business Media, Berlin, 2012)
15. M. Obrovac, V. Chevrier, Alloy negative electrodes for Li-ion batteries. Chem. Rev. **114**(23), 11444–11502 (2014)

16. C.M. Hayner, X. Zhao, H.H. Kung, Materials for rechargeable lithium-ion batteries. Annu. Rev. Chem. Biomol. Eng. **3**, 445–471 (2012)
17. B.L. Ellis, K.T. Lee, L.F. Nazar, Positive electrode materials for Li-ion and Li-batteries†. Chem. Mater. **22**(3), 691–714 (2010)
18. R. Marom et al., A review of advanced and practical lithium battery materials. J. Mater. Chem. **21**(27), 9938–9954 (2011)
19. A. Reddy, *Modern Electrochemistry 2B Electrodics in Chemistry, Engineering, Biology, and Environmental Science* (Kluwer Academic/Plenum Publishers, New York, 2001), p. 1661
20. C.D. Rahn, C.-Y. Wang, *Battery Systems Engineering* (Wiley, Hoboken, 2013)
21. J.O.M. Bockris, A.K.N. Reddy, M. Gamboa-Aldeco, *Modern Electrochemistry 2A Fundamentals of Electrodics* (Kluwer Academic/Plenum Publishers, New York, 2000)
22. D.M. Bernardi, R. Chandrasekaran, J.Y. Go, Solid-state transport of lithium in lithium-ion-battery positive electrodes. J. Electrochem. Soc. **160**(9), A1430–A1441 (2013)
23. D.R. Baker, M.W. Verbrugge, Intercalate diffusion in multiphase electrode materials and application to lithiated graphite. J. Electrochem. Soc. **159**(8), A1341–A1350 (2012)
24. K.G. Gallagher et al., A volume averaged approach to the numerical modeling of phase-transition intercalation electrodes presented for LixC6. J. Electrochem. Soc. **159**(12), A2029–A2037 (2012)
25. D.M. Bernardi, J.-Y. Go, Analysis of pulse and relaxation behavior in lithium-ion batteries. J. Power Sources **196**(1), 412–427 (2011)
26. V. Srinivasan, J. Newman, Discharge model for the lithium iron-phosphate electrode. J. Electrochem. Soc. **151**(10), A1517–A1529 (2004)
27. Y. Ji, Y. Zhang, C.-Y. Wang, Li-ion cell operation at low temperatures. J. Electrochem. Soc. **160**(4), A636–A649 (2013)
28. A. Awarke, S. Pischinger, J. Ogrzewalla, Pseudo 3D modeling and analysis of the SEI growth distribution in large format Li-ion polymer pouch cells. J. Electrochem. Soc. **160**(1), A172–A181 (2013)
29. L.O. Valøen, J.N. Reimers, Transport properties of LiPF6-based Li-ion battery electrolytes. J. Electrochem. Soc. **152**(5), A882–A891 (2005)
30. C.-F. Chen, A. Verma, P.P. Mukherjee, Probing the role of electrode microstructure in the lithium-ion battery thermal behavior. J. Electrochem. Soc. **164**(11), E3146–E3158 (2017)
31. P. Peng, Y. Sun, F. Jiang, Thermal analyses of LiCoO$_2$ lithium-ion battery during oven tests. Heat Mass Transf. **50**(10), 1405–1416 (2014)
32. C.F. Lopez, J.A. Jeevarajan, P.P. Mukherjee, Characterization of lithium-ion battery thermal abuse behavior using experimental and computational analysis. J. Electrochem. Soc. **162**(10), A2163–A2173 (2015)
33. C.F. Lopez, J.A. Jeevarajan, P.P. Mukherjee, Experimental analysis of thermal runaway and propagation in lithium-ion battery modules. J. Electrochem. Soc. **162**(9), A1905–A1915 (2015)
34. T. Hatchard et al., Thermal model of cylindrical and prismatic lithium-ion cells. J. Electrochem. Soc. **148**(7), A755–A761 (2001)
35. P.P. Mukherjee, S. Pannala, J.A. Turner, *Modeling and Simulation of Battery Systems. Handbook of Battery Materials*, 2nd edn. (2011) pp. 841–875
36. A. Mistry et al., Analysis of long-range interaction in lithium-ion battery electrodes. J. Electrochem. Energy Convers. Storage **13**(3), (2016)
37. B. Vijayaraghavan et al., An analytical method to determine tortuosity in rechargeable battery electrodes. J. Electrochem. Soc. **159**(5), A548–A552 (2012)
38. M. Ebner, V. Wood, Tool for tortuosity estimation in lithium ion battery porous electrodes. J. Electrochem. Soc. **162**(2), A3064–A3070 (2015)
39. M. Ebner, et al., Tortuosity anisotropy in lithium-ion battery electrodes. Adv. Energy Mater. **4**(5) (2014)
40. V.D. Bruggeman, Berechnung verschiedener physikalischer Konstanten von heterogenen Substanzen. I. Dielektrizitätskonstanten und Leitfähigkeiten der Mischkörper aus isotropen Substanzen. Ann. Phys. **416**(7), 636–664 (1935)

Chapter 5
Porous Media Applications: Biological Systems

Malay K. Das and Chandan Paul

5.1 Introduction

In biological systems and biomedical devices, the flow of mass and energy often involve transport through natural and artificial porous media. Examples include water, minerals, and nutrient flow in plants through xylem and phloem as well as transport in extracellular space in the central nervous system of animals. Porous media applications also find place in tissue engineering where morphogenesis and pathogenesis in regenerating tissues are greatly influenced by the fluid transport through porous scaffolds. In many biological applications, the dynamics of blood flow through porous media play crucial role in the physiological processes [1]. One such phenomenon is the coil embolization treatment of aortic and cerebral aneurysm has recently received considerable attention.

5.1.1 Aneurysm and the Treatment Options

A cerebral aneurysm is a cerebrovascular disorder in which weakness in the wall of a cerebral artery or vein causes a localized dilation or ballooning of the blood vessel. A common location of cerebral aneurysms is on the arteries at the base of the brain, known as the Circle of Willis. Cerebral aneurysms can leak or rupture causing symptoms from severe headache to stroke-like symptoms and, if untreated, may lead to death. Similarly, abdominal aortic aneurysm is a localized dilatation of the abdominal aorta exceeding the normal diameter by more than 50% and is the most common form of aortic aneurysm. Smoking, alcoholism, and hypertension are among the common causes of aortic aneurysm. If left untreated, such aneurysm may lead to serious complications leading to death. There are, however, several treatment options available for aortic and cerebral aneurysm [2].

© Springer International Publishing AG 2018
M.K. Das et al., *Modeling Transport Phenomena in Porous Media with Applications*, Mechanical Engineering Series,
https://doi.org/ 10.1007/978-3-319-69866-3_5

While surgery and stent grafting constitute the prominent treatment option for the aortic aneurysm, endovascular coil embolization is a favorable alternative for cerebral aneurysm [3]. In endovascular coil embolization or coiling, the surgeon inserts a catheter into a blood vessel and using X-ray tracking, the catheter is guided to the correct aneurysm position. Once in position, the surgeon inserts tiny metal coils inside the aneurysm. The softness of the coils allows the coil to conform to the often irregular shape of an aneurysm. An average of 5–6 coils is required to completely pack an aneurysm. The coils cause the aneurysm to clot off and prevent rupture. Aneurysms with a wide neck or unusual shape require a stent to help hold the coils in place. The risk due to embolization is low. Possible complications include stroke-like symptoms such as weakness in one arm or leg, numbness, tingling, speech disturbances, and visual problems. Serious complications such as permanent stroke or death are rare.

5.1.2   Blood Flow in Coil-Embolized Aneurysm: Role of Modeling and Simulation

Yu and Zhao [4], Lieber et al. [5], Liou et al. [6], and Canton et al. [7] have experimentally used an idealized geometry to study the influence of stents on intra-aneurysmal flow. Gobin et al. [8] and Canton et al. [9] reported the effect of coil embolization on an idealized intracranial aneurysm. Ford et al. [10] and Sun et al. [11] conducted experiments to analyze the flow dynamics in intracranial aneurysm with patient-specific geometry. Goubergrits et al. [12] experimentally investigated the effect of coil embolization in a cerebral aneurysm under steady flow conditions.

While laboratory experiments suffer from the difficulties of geometric reconstruction of actual aneurysm, several models and simulations are also being developed. Aenis et al. [13] and Ohta et al. [14] reported a finite element study of sidewall aneurysm using idealized geometry. Stuhne and Steinman [15] conducted a numerical study to analyze wall shear distribution on a stented sidewall aneurysm using a model geometry. Steinman et al. [16] presented image-based computational simulations of the flow dynamics in an anatomically realistic cerebral aneurysm assuming rigid walls. Rayz et al. [17] carried out computational fluid dynamic (CFD) simulation using patient-specific geometry for predicting the region prone to thrombus formation. Rayz et al. [18] conducted another numerical study to compare CFD results with in vivo MRI measurements. Takizawa et al. [19, 20] reported numerical simulation of blood flow through stented cerebral aneurysm considering fluid–structure interaction model. Valencia et al. [21] studied blood flow dynamic in several patient-specific cerebral aneurysm models. They presented the relationship between wall shear stress and aneurysm area index.

Very few studies are available in the literature on numerical simulation of flow in a tube with coil embolization. Byun and Rhee [22] modeled the coils as a small solid sphere placed at different location within the aneurysm. Groden et al. [23] studied three-dimensional pulsatile flow in cerebral aneurysm before and after coil embolization. These authors used cube-shaped cells to represent the coils. Cha et al. [24] modeled coil embolization with porous medium in an idealized geometry. The authors used circular tubes of the surrounding arteries with a spherical shape of the aneurysm. Later Khanafer et al. [25] presented a numerical model to quantify the reduction in blood velocity and pressure resulting from the placement of endovascular coils using volume-averaged porous medium equation. Recently, Otani et al. [26] have studied the effect of packing density of coils on flow stagnation on an idealized geometry using a porous medium model. Mitsos et al. [27] presented a steady flow simulation in patient-specific cerebral aneurysm assuming coils to be a porous medium. In recent years, Paul et al. [28] and Agarwal et al. [29] presented CFD simulation of coil-embolized aneurysm in idealized geometries.

5.2 Analytical Solutions of Flow in a Channel and a Tube

Analytical solutions of fully developed steady and pulsatile flow in clear and porous media form a classical problem in fluid mechanics. The parabolic velocity profile in a clear fluid for both a channel and a tube is well-known as the Poiseuille flow solution. Some other important solutions are first surveyed below:

5.2.1 Pulsatile Flow Through a Tube with Clear Media

Womersley [30] and Sazonov et al. [31] analytically studied laminar, fully developed pipe flow under oscillatory conditions. The tube is taken to be filled with clear fluid. The governing equation for such a geometry is given as:

$$\frac{\partial w}{\partial t} = -\frac{\partial p}{\partial z} + \frac{1}{Re}\left(\frac{\partial^2 w}{\partial r^2} + \frac{1}{r}\frac{\partial w}{\partial r}\right) \tag{5.1}$$

Here, w is the local fluid velocity component in the axial direction (z) while the other two velocity components, under the fully developed approximation, are zero. In addition, r is the radial coordinate scaled by the diameter of the tube. Further,

$$w = w(r, t); \ p = p(z, t) \tag{5.2}$$

For fully developed flow, the pressure gradient is constant in space though it is time-dependent. Accordingly, pressure varies linearly in the axial direction. With a velocity scale U and pipe diameter d, Reynolds number (Re) is defined in the present context as:

$$\text{Re} = \frac{\rho U d}{\mu} \tag{5.3}$$

With the symbol ω as the fundamental frequency of the pulsatile flow, Womersley [30] suggested the following forms of axial velocity and pressure gradient:

$$w(r,\, t) = \text{real}\left[\sum_{n=-\infty}^{\infty} W_n(r)\, e^{i\omega_n t}\right]; \; \omega_n = n\omega \;\; \text{and}$$

$$-\frac{\partial p}{\partial z}(t) = -\text{real}\left[\sum_{n=-\infty}^{\infty} G_n\, e^{i\omega_n t}\right] \tag{5.4}$$

Substituting Eq. 5.4 in 5.1 and solving the resulting ordinary differential equation, the following solution can be derived [29]:

$$W_n(r) = W_{cn}\left(1 - \frac{I_0\left(r\sqrt{A_n}\right)}{I_0\left(0.5\sqrt{A_n}\right)}\right)\bigg/\left(1 - \frac{1}{I_0\left(0.5\sqrt{A_n}\right)}\right); \; n \neq 0$$

and $W_n(r) = W_{cn}\left(1 - 4r^2\right); \; n = 0 \tag{5.5}$

with $A_n = i\text{Re}\omega_n$

Further, W_{cn} is the n-th harmonic of the centerline velocity and I_j is the modified Bessel function of the j-th order. It can be shown that:

$$W_{cn} = B_n\left(1 - \frac{1}{I_0\left(0.5\sqrt{A_n}\right)}\right); \; n \neq 0$$

$$\text{and } W_{cn} = \frac{G_n}{16\text{Re}}; \; n = 0 \tag{5.6}$$

$$\text{where } B_n = \frac{G_n}{i\omega_n}$$

The above analytical solution, known as the *Womersley solution* may be utilized in the following two ways:

(i) If the pressure gradient waveform $\partial p/\partial z$ is known, then it can be decomposed into Fourier components as shown in Eq. 5.4 and coefficients G_n can be found. Now, the centerline velocity for the n-th harmonic W_{cn} can be calculated with the help of Eq. 5.6. Thereafter the final velocity profile may be obtained by referring to Eq. 5.4.

(ii) If centerline velocity waveform $W_c(t)$ is known (or, equivalently the waveform of the volume flow rate), it can be decomposed into Fourier components, evaluating coefficients W_{cn} as shown in Eq. 5.4. With the known harmonics W_{cn}, velocity profiles can be calculated using Eq. 5.5.

Typical flow patterns arising in a clear tube and one-filled with a porous medium are graphically portrayed in Sect. 5.3.

5.2.2 Steady Flow Through a Channel Filled with Porous Media

The governing equation for laminar fully developed flow in a parallel plate channel is given in dimensionless form as:

$$\frac{dp}{dz} = \frac{1}{\varepsilon \, \text{Re}} \left(\frac{\partial^2 w}{\partial y^2} \right) - \frac{w}{\text{ReDa}} \tag{5.7}$$

Here ε is porosity, K is permeability, d is the channel opening, and the Darcy number (Da) is defined as:

$$\text{Da} = \frac{K}{d^2} \tag{5.8}$$

Symbol w is the velocity in the streamwise (z) direction and y indicates the spanwise direction across the channel width. Referring to the non-Darcy equation of Chap. 2, it is clear that Eq. 5.7 is a linearized form that neglects the Forchheimer term. Equivalently, it is the Brinkman form of the momentum equation for steady flow in a porous medium. Enforcing the no-slip condition for velocity at $y = \pm 0.5$ and utilizing the fully developed flow approximation, namely, $w = w\,(y)$, $p = p(z)$, the following solution is obtained:

$$w(y) = w_c \left[1 - \frac{\cosh\left(y\sqrt{A} \right)}{\cosh\left(0.5\sqrt{A} \right)} \right] \Bigg/ \left[1 - \frac{1}{\cosh\left(0.5\sqrt{A} \right)} \right] \tag{5.9}$$

$$w_c : \text{ centerline velocity and } A = \frac{\varepsilon}{\text{Da}}$$

The centerline velocity may be calculated either from the prescribed mass flow rate or the stipulated pressure gradient, as practiced in the context of flow of fluids in a clear media.

5.2.3 Steady Flow Through a Tube Filled with Porous Media

The solution obtained for flow in a channel is now extended to flow through a tube. The Brinkman equation for laminar, fully developed case steady flow in a tube is given as:

$$\frac{dp}{dz} = \frac{1}{\varepsilon \, \text{Re}} \left(\frac{\partial^2 w}{\partial r^2} + \frac{1}{r} \frac{\partial w}{\partial r} \right) - \frac{w}{\text{ReDa}} \tag{5.10}$$

where w is the velocity in streamwise direction (z) and r indicates the radial direction scaled by the tube diameter. Here, the governing equation is linearized by neglecting the Forchheimer term of the full non-Darcy model (Chap. 3). Further, by enforcing the no-slip condition at the wall, $r = 0.5$ and utilizing the conditions of fully developed flow, namely, $w = w \ (r)$, $p = p(z)$, the following solution is obtained:

$$w(r) = w_c \left[1 - \frac{I_0\left(r\sqrt{A}\right)}{I_0\left(0.5\sqrt{A}\right)} \right] \Bigg/ \left[1 - \frac{1}{I_0\left(0.5\sqrt{A}\right)} \right] \tag{5.11}$$

$$\text{with } w_c : \text{ centerline velocity and } A = \frac{\varepsilon}{\text{Da}}$$

Symbol I_0 is the modified Bessel function of 0th order. The above equation may be stated in terms of average velocity as follows:

$$\frac{w}{w_{avg}} = \left(1 - \frac{I_0\left(r\sqrt{A}\right)}{I_0\left(0.5\sqrt{A}\right)} \right) \Bigg/ \left(1 - \frac{4}{\sqrt{A}} \frac{I_1\left(r\sqrt{A}\right)}{I_0\left(0.5\sqrt{A}\right)} \right) \tag{5.12}$$

$$w_{avg} : \text{ averge velocity over the tube cross section}$$

Here, I_1 is the modified Bessel function of first order.

5.2.4 Pulsatile Flow Through a Channel Filled with Porous Media

The Brinkman form of the governing equation for such a configuration is given as:

$$\frac{\partial w}{\partial t} = -\varepsilon \frac{\partial p}{\partial z} + \frac{1}{\text{Re}} \frac{\partial^2 w}{\partial y^2} - \frac{\varepsilon w}{\text{ReDa}} \tag{5.13}$$

Symbols such as ε, Re, and Da are as defined in Sect. 5.2.2. Equation 5.13 is the linearized form of the full non-Darcy model since it neglects the Forchheimer term. Flow is treated as fully developed in space though pulsatile in time. Accordingly, symbol w, the velocity component in the axial (z) direction is spatially a function of the y-coordinate alone while the other two velocity components, under the fully developed flow approximation, are zero. In addition, pressure gradient is a constant in space though pulsatile in time. Hence,

$$w = w(y, t); \; p = p(z, t) \tag{5.14}$$

Using Fourier series expansion with respect to the fundamental frequency ω, the axial velocity and pressure gradient are expressed as:

$$w = \text{real}\left[\sum_{n=-\infty}^{\infty} W_n(y)e^{i\omega_n t}\right]; \; \omega_n = n\omega;$$

$$-\frac{\partial p}{\partial z} = \text{real}\left[\sum_{n=-\infty}^{\infty} G_n e^{i\omega_n t}\right] \tag{5.15}$$

Substituting Eq. 5.15 in 5.14 and solving the resulting ordinary differential equation, we get a solution for the velocity amplitude as:

$$W_n(y) = W_{cn}\left(1 - \frac{\cosh\left(r\sqrt{A_n}\right)}{\cosh\left(0.5\sqrt{A_n}\right)}\right) \Big/ \left(1 - \frac{1}{\cosh\left(0.5\sqrt{A_n}\right)}\right)$$

$$W_n(y = 0.5) = W_{cn} = B_n\left(1 - \frac{1}{\cosh\left(0.5\sqrt{A_n}\right)}\right) \tag{5.16}$$

$$\text{where } A_n = i\text{Re}\omega_n + \frac{\varepsilon}{\text{Da}} \text{ and } B_n = \frac{G_n}{i\omega_n + \frac{\varepsilon}{\text{ReDa}}}$$

As discussed in Sect. 5.2.1, the pulsatile pressure waveform or the mass flow rate needs to be prescribed so that the constant of integration in the form of the centerline velocity in Eq. 5.16 can be determined.

5.2.5   Pulsatile Flow Through a Tube Filled with Porous Media

The discussion of Sect. 5.2.4 is extended from channel flow to one in a circular tube. The Brinkman equation for pulsatile but spatially fully developed flow in a tube for such geometry may be written as:

$$\frac{\partial w}{\partial t} = -\varepsilon\frac{\partial p}{\partial z} + \frac{1}{\text{Re}}\frac{\partial^2 w}{\partial y^2} - \frac{\varepsilon w}{\text{ReDa}} \tag{5.17}$$

Equation 5.17 is the linearized form of the non-Darcy model since the Forchheimer term is neglected. Here, w is the Darcian velocity component in the axial (z) direction and is a function of the radial coordinate and time. Under the fully developed flow approximation, the other two velocity components are zero and pressure gradient is spatially constant but a pulsatile function of time. Specifically,

$$w = w(r, \ t); \ p = p(z, \ t) \tag{5.18}$$

with ω as the fundamental frequency of pulsatile flow, the axial velocity and pressure gradient are assumed as:

$$w = \text{real}\left[\sum_{n=-\infty}^{\infty} W_n(r)\, e^{i\omega_n t}\right]; \ \omega_n = n\omega$$

$$-\frac{\partial p}{\partial z} = \text{real}\left[\sum_{n=-\infty}^{\infty} G_n e^{i\omega_n t}\right] \tag{5.19}$$

Substituting the Eq. 5.19 in 5.18 and solving the resulting ordinary differential equation, we get

$$W_n(r) = W_{cn}\left(1 - \frac{I_0\left(r\sqrt{A_n}\right)}{I_0\left(0.5\sqrt{A_n}\right)}\right) \Big/ \left(1 - \frac{1}{I_0\left(0.5\sqrt{A_n}\right)}\right)$$

$$W_n(y = 0.5) = W_{cn} = B_n\left(1 - \frac{1}{I_0\left(0.5\sqrt{A_n}\right)}\right) \tag{5.20}$$

$$\text{where } A_n = i\text{Re}\omega_n + \frac{\varepsilon}{\text{Da}} \ \text{ and } \ B_n = \frac{G_n}{i\omega_n + \frac{\varepsilon}{\text{ReDa}}}$$

To utilize the above analytical solution, pulsatile pressure waveform or the mass flow rate should be prescribed, as discussed in Sect. 5.2.1.

5.3   Pulsatile Flow in a Porous Bulge

In this section, we discuss a generic model of pulsatile flow through porous media [32–39]. Nonlinear terms are retained, flow is three-dimensional and unsteady, and the geometry is not a channel or a tube. Analytical solutions are not possible, and a numerical route is recommended for solving the system of equations. Since flows such as cardiovascular and pulmonary circulation are oscillatory in nature, a variety of biological systems are modeled as pulsatile flow through porous media [40–42]. Examples include clinical atherosclerosis, in which blood flows through the arteries that are partially clogged due to deposition of porous, fatty materials such as cholesterol and triglyceride [43–46]. Similarly, transport in soft tissues sarcoma and blood vessel tumors may also be modeled as pulsatile flow through complex, disorganized, biological porous media [47–50].

In the present simulation, the geometry under investigation comprises a circular tube of diameter d with an axisymmetric bulge in the middle (Fig. 5.1). The bulge diameter varies from d to a maximum of D over a length L as per the following geometric relations [51]:

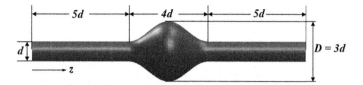

Fig. 5.1 Schematic diagram and major dimensions of a tubular geometry with a bulge. The bulge serves as an idealized setting for fluid flow simulation through a coil-embolized aneurysm

$$g(z) = d \quad \text{for } z \leq 5d \text{ and } z \geq 9d \tag{5.21}$$

$$g(z) = \frac{D-d}{2}\left[1 + \sin\left\{\frac{2\pi(z-5d)}{L} - \frac{p}{2}\right\}\right] + d \tag{5.22}$$

$$\text{for } 5d < z < 9d \text{ where } D = 3d, \ L = 4d$$

A bulge in an artery is a deformity called an *aneurysm* in the medical literature. Progressive growth of the bulge can lead to wall thinning and rupture, a consequence that can result in fatality. A possible medical treatment that can arrest further growth of the bulge is the placement of a metallic coil in the ballooned port of the artery. Flow in the tube with a coil can be simulated by modeling the flow space as a porous medium. This approach is further developed below.

5.3.1 Governing Equations

The fluid, blood in the biomedical context, is considered Newtonian and incompressible for the present discussion while the flow is laminar and unsteady. The porous medium is homogeneous, isotropic, and saturated with the flowing fluid. The dimensionless form of the non-Darcy equations of mass and momentum for the fluid in the porous medium is given as follows [Chap. 2; also see 52–56]:

$$\nabla \cdot \mathbf{u} = 0 \tag{5.23}$$

$$\varepsilon^{-1}\frac{\partial \mathbf{u}}{\partial \tau} + \varepsilon^{-2}(\mathbf{u} \cdot \nabla)\mathbf{u} = -\nabla p + \frac{\nabla^2 \mathbf{u}}{\varepsilon \mathrm{Re}} - \frac{\mathbf{u}}{\mathrm{ReDa}} - \frac{C_f}{\sqrt{\mathrm{Da}}}[\mathbf{u} \cdot \mathbf{u}]J \tag{5.24}$$

$$\text{For a clear medium: } \varepsilon = 1; \ K \to \infty; \ \mathrm{Da} \to \infty \tag{5.25}$$

Symbol **u** is the Darcian velocity and J is the unit vector in the flow direction. Other symbols appearing in Eqs. 5.23–5.25 are as explained in Sect. 5.2.2. The reference quantities used for the non-dimensionalization of the governing equations are given in Table 5.1.

To account for pulsation in the mass flow rate, the inlet velocity profile is decomposed into temporal and spatial components. While the temporal component

Table 5.1 Reference variables used for non-dimensionalization

Length	d (tube diameter)
Velocity	U (reference velocity)
Time	d/U
Pressure	ρU^2
Wall shear stress (WSS)	$\mu U/d$
Angular frequency	U/d
Vorticity	U/d

uses reported experimental results, the spatial component of the inlet velocity profile is obtained via solution of the steady linearized governing equations. For the clear medium, the Hagen–Poiseuille solution specifies a parabolic variation in space, as shown in Eq. 5.26. Similarly, for the porous medium, the analytical velocity profile is given by Eq. 5.27 and is derived by dropping the nonlinear inertia and Forchheimer terms Eq. 5.24. The inflow boundary conditions are summarized as follows:

$$\text{Clear medium, inlet: } z = 0, \ w(r, \ t) = 2w_{\text{avg}}(t)\left(1 - 4r^2\right) \quad (5.26)$$

Porous medium, inlet: $z = 0$,

$$w(r, \ t) = w_{\text{avg}}(t)\left[1 - \frac{I_0\left(r\sqrt{A}\right)}{I_0\left(0.5\sqrt{A}\right)}\right] \Big/ \left[1 - \frac{4}{\sqrt{A}}\frac{I_1\left(r\sqrt{A}\right)}{I_0\left(0.5\sqrt{A}\right)}\right] \quad (5.27)$$

Here, $A = \varepsilon/\text{Da}$, while I_0 and I_1 are modified Bessel functions of zeroth order and first order, respectively. The function $w_{\text{avg}}(t)$ represents the variation of the average velocity with time at the inflow plane of the modeled artery. For the present simulation, an experimentally obtained velocity waveform has been used [52–55]. With this data, $w_{\text{avg}}(t)$ is approximated as a truncated Fourier series, Eq. 5.28 and plotted in Fig. 5.2.

$$\begin{aligned} w_{\text{avg}}(t) = 1.405[&0.06327 + 0.21008\sin(2\pi\tau) + 0.21651\cos(2\pi\tau) \\ &+ 0.22533\sin(4\pi\tau) + 0.0064\cos(4\pi\tau) \quad (5.28) \\ &+ 0.05807\sin(6\pi\tau) - 0.11105\cos(6\pi\tau)] \end{aligned}$$

$$\tau = \frac{td}{UT}; \ T, \ \text{time-period of pulsatile flow} \quad (5.29)$$

5.3.2   Flow Parameters

In the present chapter, data is reported for simulations conducted at Re = 500, 2000 and Da = 10^{-2}, 10^{-4}. The above parameters indicate high and low inertia as well as

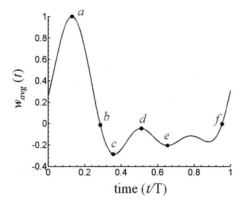

Fig. 5.2 Non-dimensional average velocity variation with time at the inflow plane, resembling a vascular flow rate. Phases (a)–(f) indicate time instants for which the solution has been presented. The peak phase for velocity is 'a,' the zero phases are 'b' and 'f,' while 'c,' 'd,' and 'e' are other peaks of the waveform. The peak velocity is unity indicating that the Reynolds number is based on the velocity at the peak phase, namely 'a'

high and low permeability, respectively. Further, to represent the pulsatile nature of flow, we use the Womersley number (Wo) defined as follows [56–58]:

$$\text{Wo} = \frac{d}{2}\sqrt{\frac{2\pi f}{v}}; f : \text{frequency of oscillation in Hz} \qquad (5.30)$$

Simulations are conducted for $f = 1.2$ Hz leading to Wo = 11.3 for the geometric data of Fig. 5.1. Here, data related to the effective density (1157 kg/m^3) and viscosity of blood (9.2 × 10^{-3} Pa-s), treated as a single-phase continuum is used. The values of Reynolds and Womersley numbers thus derived are quite suitable for biomedical simulation. The detail description of the flow parameters used in the current simulations is presented in Table 5.2.

Table 5.2 Flow parameters used for computation of pulsatile flow through a tube with bulge

Comparison 1	Re = 2000 $f = 1.2$ Hz Wo = 11.3	Clear medium
		Porous medium (high permeability): $\varphi = 0.85$, Da = 10^{-2}, $C_f = 0.2$
		Porous medium (low permeability): $\varphi = 0.6$, Da = 10^{-4}, $C_f = 0.3$
Comparison 2	Re = 500 $f = 1.2$ Hz Wo = 11.3	Clear medium
		Porous medium (high permeability): $\varphi = 0.85$, Da = 10^{-2}, $C_f = 0.2$
		Porous medium (low permeability): $\varphi = 0.6$, Da = 10^{-4}, $C_f = 0.3$

5.3.3 Pulsatile Flow in a Porous Bulge: Numerical Solution

The system of Eqs. (5.23–5.25) is quite involved and cannot be solved analytically. Here, an unstructured finite-volume based, parallelized CFD solver is used to solve for the flow and pressure fields in the porous bulge [59] using the non-Darcy equations of Chap. 2. The code has been extensively validated against analytical, numerical, and experimental data, both for steady as well as unsteady flow fields. While the computational grid is generated using commercial software ICEM-CFD®, the finite volume solver has been developed with particular emphasis on multiphase and porous media applications. The nature of grids is shown in Fig. 5.3. Figures 5.3a–c, respectively, show unstructured grids in the (a) tube cross section, (b) bulge cross section, and (c) the side view of the grids. The bulge geometry of interest (Fig. 5.1) represents the idealized form of coil-embolized aneurysm. The related flow, pressure, and shear stress distributions at each instant of time of the pulsatile waveform are quantities of interest.

5.3.4 Validation of the Finite Volume Solver

The geometries considered in the present section may facilitate simplification of the governing equations in terms of dimensionality. This advantage is utilized for deriving the analytical solution. The numerical solver of the non-Darcy equations is, however, fully unsteady and three-dimensional, though the solution is obtained in a Cartesian coordinate system with Cartesian velocity components. The details of the configurations considered for validation of steady flow are given below:

(i)   Numerical solution of steady flow through a tube filled with porous medium (without the Forchheimer term) is validated against the analytical solution for two Darcy numbers. At the inlet, uniform velocity profile is given as the boundary condition to the FVM solver and flow is allowed to develop in the axial direction. The analytical solution, described in Sect. 5.1, is used to validate the numerical solution. Figure 5.4a shows that the numerical velocity profile matches the analytical solution exceedingly well. In addition, a thinning of the boundary layer with increasing Reynolds number is clearly revealed.

(ii) Numerical solution of steady flow through a parallel plate channel filled with porous medium (without the Forchheimer term) is validated against the analytical solution. A flat velocity profile is given as inlet boundary condition. In this case, the analytical velocity profile for fully developed flow is obtained, as described in Sect. 5.1. Figure 5.4b depicts a close match between the numerical and analytical velocity profiles for two different Darcy and Reynolds numbers.

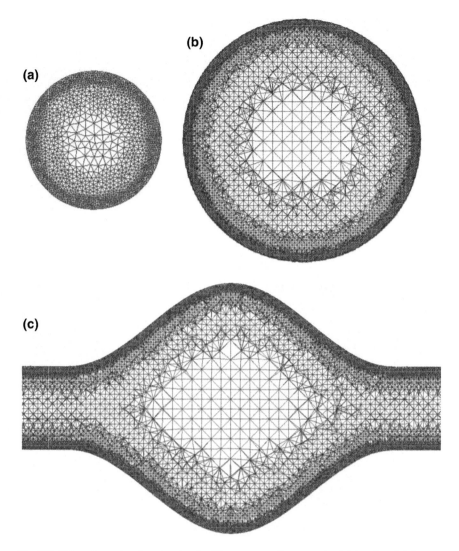

Fig. 5.3 Three-dimensional unstructured grid: **a** cross section near the inlet and outlet, **b** cross section at the bulge, **c** as seen from the side of the bulge section

(iii) The governing equation for fully developed steady flow through a tube filled with porous medium the includes the Forchheimer term is given as:

$$\frac{dp}{dz} = \frac{1}{\varepsilon \, \text{Re}} \left(\frac{\partial^2 w}{\partial r^2} + \frac{\partial w}{r \partial r} \right) - \frac{w}{\text{ReDa}} - \frac{C_f}{\sqrt{\text{Da}}} |w| w \qquad (5.31)$$

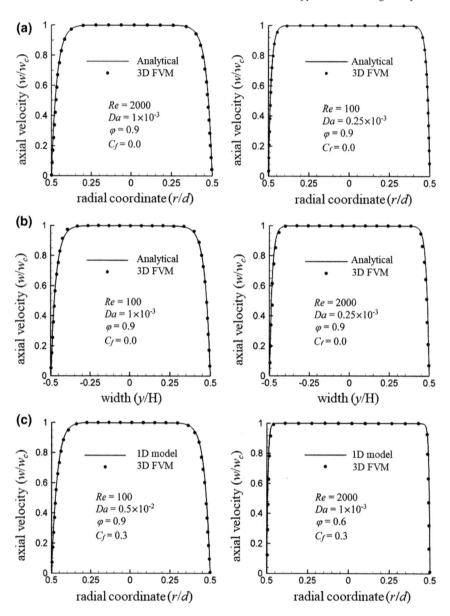

Fig. 5.4 Validation of the FVM solver for steady flow in a porous medium against an analytical solution and from an ODE solver: **a** tube and **b** channel geometry; **c** tube including Forchheimer term

Here, dp/dz = constant. A compact analytical solution of the differential Eq. 5.31 is not possible. However, with a prescribed pressure gradient, Eq. 5.31 can be solved using standard ordinary differential equation solvers. This solution is compared against that generated from the 3-D FVM solver. For FVM, the inlet prescription is a flat velocity profile and at the outlet, a known pressure boundary condition is specified. The pressure gradient obtained from the FVM solver is used to solve Eq. 5.31 as an ODE, with no-slip boundary condition at the wall, $r = 0.5$. Figure 5.4c shows the comparison between the 3-D numerical solver and the 1-D solution. The two velocity profiles show an excellent match.

The following case studies are considered for validating the 3-D FVM solver against the analytical solutions under pulsatile flow conditions:

(i) Pulsatile flow through a tube with clear fluid is validated against the Womersley solution (Sect. 5.1). To ensure fully developed flow in the tube, its length-to-diameter ratio is taken as 60. At the inlet of the tube, a sinusoidal waveform of the form $w(r,t) = (1 - 4r^2)\sin(2\pi t)$ is given as an inflow boundary condition. At the outlet, a known (datum) pressure is specified. Here, r and t are the non-dimensional radial coordinate scaled by the diameter and time scaled by the time period T. The centerline velocity waveform at the outlet (after 10 cycles) is shown in Fig. 5.5a. To get the Womersley solution, the outlet centerline velocity waveform from the numerical simulation is decomposed into 41 Fourier components (-20 to 20). The decomposed Fourier function exactly matches the analytical waveform (Fig. 5.5a) except near the boundary points. With the help of the Fourier-decomposed function, the analytical velocity profiles are calculated at different time instances, as described in Sect. 5.1. Figure 5.5c shows that the numerical velocity profiles closely match the Womersley solution at the phases considered. The velocity overshoot characteristic of a Womersley profile is brought out well by the numerical solution.

(ii) Pulsatile flow through a tube filled with homogeneous isotropic porous medium (without the Forchheimer term) is validated against the analytical solution (Sect. 5.1). For deriving a numerical solution, the inlet is provided with the spatiotemporal velocity distribution of the form $w(r,t) = (1 - 4r^2)\sin(2\pi t)$. At the outlet, a known (datum) pressure is specified. The centerline velocity waveform at the outlet is used for comparison with the analytical solution shows the comparison between the analytical and numerical velocity profiles at specified time instants. From the figure, it is evident that the numerical profile closely matches the analytical profile. The role of the porous medium is seen to flatten the velocity distribution across the tube.

(iii) Pulsatile flow through a parallel plate channel filled with porous medium is validated against the analytical solution (Sect. 5.1). At the inlet of the channel, a sinusoidal waveform of the form $w(r,t) = (1 - 4y^2)\sin(2\pi t)$ is provided. At the outlet, a known (datum) pressure is specified. Using the centerline velocity waveform at the outlet (Fig. 5.5c), the analytical velocity

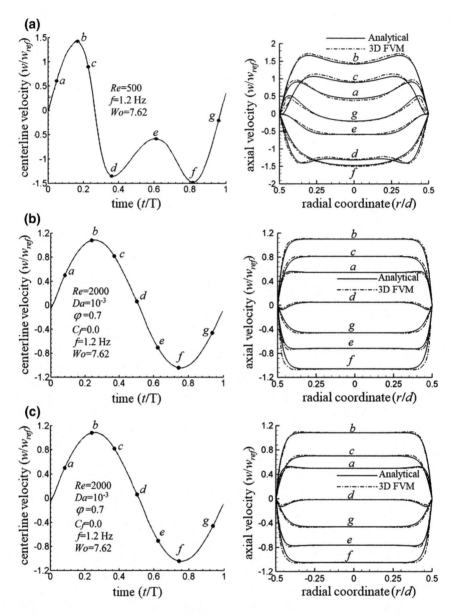

Fig. 5.5 Validation of the numerical solver against the analytical for oscillatory flow in porous and clear media. **a** Tube and **b** tube filled with porous medium **c** Channel filled with porous medium

profile for fully developed flow is calculated on the outflow plane, as described in Sect. 5.1 shows a close match between the analytical and numerical velocity profiles. In the discussion above, W_{ref} represents the average velocity at the peak phase.

The validation of the numerical solution under oscillatory flow conditions is important since it accounts for flow reversal at the outflow plane. In a biomedical context, the pulsatile waveform has a finite time-averaged flow through the arterial vessel though the profile is time-periodic. In a numerical context, the finite outflow at any instant of time provides for a simplification and the numerical solutions are expected to be superior to purely oscillatory flow.

5.3.5   Pulsatile Flow in a Bulge

Following the validation studies of Sect. 5.3.4, simulations have been carried out to understand the details of pulsatile flow through the geometry of a bulge filled with porous material. The inlet space-averaged velocity profile, adapted from realistic blood flow measurements, is shown in Fig. 5.2. The figure includes phases that have been selected for displaying the flow distribution. Peak Reynolds numbers considered are 500 and 2000. The flow patterns in a porous medium filled model are compared with a model filled with clear fluid.

Figures 5.6, 5.7, 5.8, 5.9, 5.10, 5.11 show the instantaneous distribution of the field variables at selected phases of interest. In Figs. 5.6, 5.7, 5.8, 5.9, 5.10, 5.11, the axial and absolute velocities are presented at the midplane, while pressure, vorticity, and shear are at the wall. In these figures, the region of interest is the bulge itself and not the connecting tubes.

Figures 5.6, 5.7, 5.8 show numerical simulation at Re = 500. It is observed that in a porous medium, at the peak phase 'a,' a steep velocity gradient appears at the wall. At other phases, the velocity gradient varies smoothly from the wall to the center of the bulge. In a clear medium, velocity gradient varies smoothly over space for all the phases studied. In a porous medium, velocity profile at this phase is flatter across the cross section. Darcy resistance being proportional to the velocity magnitude reaches its maximum and spreads across the cross section, all the way up to the wall. Viscous effects within the fluid phase are only effective in the vicinity of the wall, appearing in the form of a steep wall velocity gradient. At the other phases, velocity magnitude is significantly smaller as compared to the peak phase. Hence, the viscous effect originating from the wall spreads toward the center. Similar conclusions can be derived from the velocity field at Re = 2000 by comparing Figs. 5.9, 5.10, 5.11. In contrast, for a clear medium at Re = 2000, the velocity contours at individual phases show fluctuations, flow separation, and asymmetry. For Re = 2000, the flow field is close to the transitional regime, which is a likely reason for the small-scale perturbations.

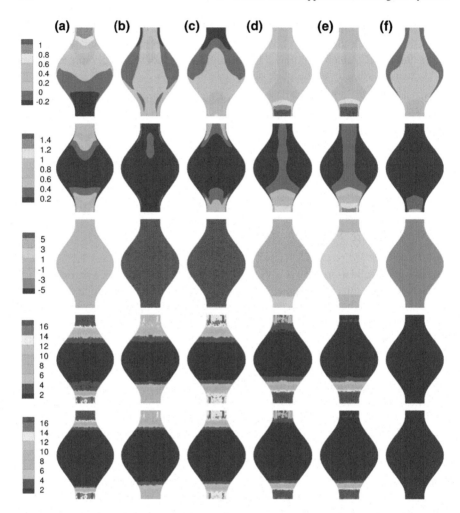

Fig. 5.6 Distribution of instantaneous axial velocity (at midplane), absolute velocity (at midplane), wall pressure, wall vorticity, and wall shear stress in a clear medium at selected phases of the cycle shown in Fig. 5.2. Re = 500 and Wo = 11.3

Comparing the instantaneous pressure data for both the Reynolds numbers, it is evident that the pressure range increases with decreasing permeability. In a porous medium, the peak pressure occurs in the bulge at the peak phase 'a'; while in a clear medium the peak pressure extends up to the phase 'b.' This happens because of the changing phase difference between pressure gradient and axial velocity profile. At the Darcian limit (when the non-Darcy model reduces to Darcy's law), pressure gradient and velocity remain correlated for every instant of time. In a clear medium, the phase difference is greater and with decreasing permeability, the phase difference decreases as we approach the Darcy limit.

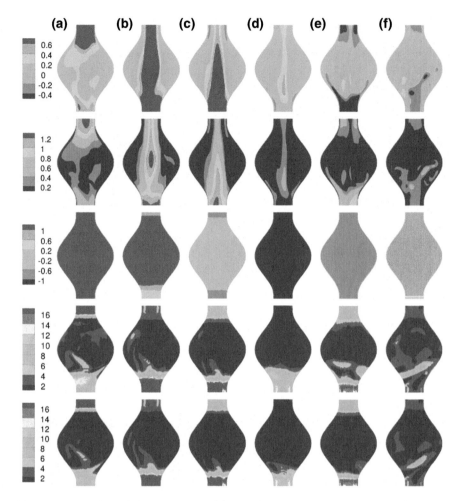

Fig. 5.7 Distribution of instantaneous axial velocity (at midplane), absolute velocity (at midplane), wall pressure, wall vorticity, and wall shear stress in clear medium at selected phases of the cycle shown in Fig. 5.2. Re = 2000, and Wo = 11.3

The wall vorticity and wall shear stress are directly dependent on the local velocity gradient. Owing to steep velocity gradients at the wall, high magnitude of wall vorticity and wall shear stress are recorded at the peak phase 'a' for both Re = 500 and 2000. At other phases, magnitudes of wall vorticity and wall shear stress depend upon permeability. For low permeability (Da = 10^{-4}), the steep velocity gradient still persist at the other phases. As a result, the vorticity and shear stress magnitude remain the highest at the wall.

Comparing the flow data in clear media from Re = 500 and Re = 2000 (Figs. 5.6 and 5.7), it is observed that the flow remains symmetric for Re = 500 but is asymmetric at Re = 2000. This is indicative of symmetry breaking phenomenon

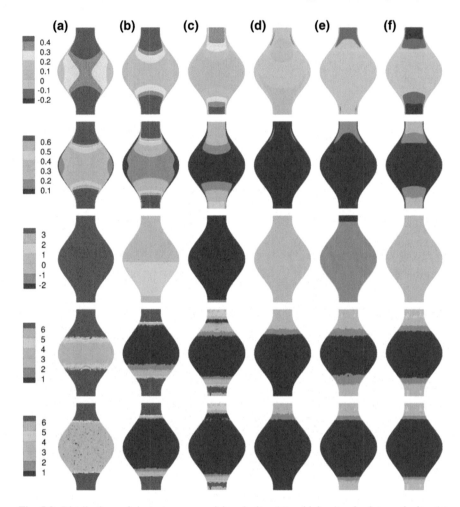

Fig. 5.8 Distribution of instantaneous axial velocity (at midplane), absolute velocity (at midplane), wall pressure, wall vorticity, and wall shear stress in porous medium at selected phases of the cycle shown in Fig. 5.2. Da = 10^{-2}, Re = 500, and Wo = 11.3

and is referred to as *pitchfork bifurcation*. In the geometry being considered, bifurcation in flow pattern occurs between Re = 500 and Re = 2000. Asymmetric velocity field leads to asymmetry in wall shear and pressure distribution, which, in the biomedical context, can lead to failure of the tissue material. In this respect, the porous medium filled bulge is superior since it equalizes pressure and wall shear stress around its periphery.

Insertion of porous media, with Da = 10^{-2}, shows two distinct effects (Figs. 5.8 and 5.9). First, the pitchfork bifurcation at Re = 2000 disappears indicating the influence of the porous media the flow structure. Secondly, as compared to the clear media, the shear stress shows less dependence on the flow Re. Further reduction of

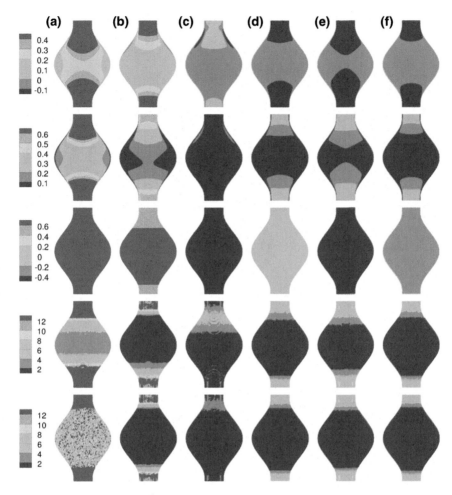

Fig. 5.9 Distribution of instantaneous axial velocity (at midplane), absolute velocity (at midplane), wall pressure, wall vorticity, and wall shear stress in porous medium at selected phases of the cycle shown in Fig. 5.2. Da = 10^{-2}, Re = 2000, and Wo = 11.3

Da (Da = 10^{-4}, Figs. 5.10 and 5.11) show highly diffusive flow patterns that are largely independent of Re. Small values of Da reduces the inertia forces leading to minimal influence of Re on the flow structure. From biomedical perspective, it may be noted that in insertion of porous media in an aneurysm reduces the effects of flow rate and flow pulsation in the aneurysm.

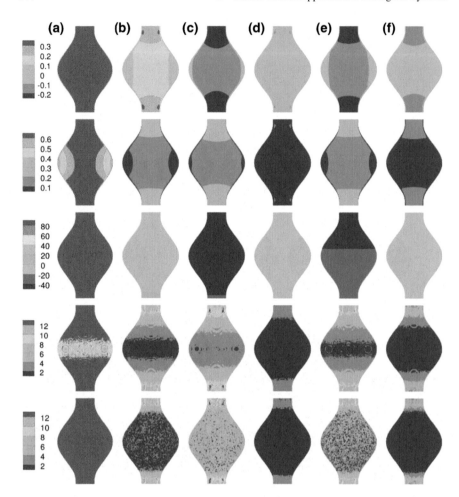

Fig. 5.10 Distribution of instantaneous axial velocity (at midplane), absolute velocity (at midplane), wall pressure, wall vorticity, and wall shear stress in porous medium at selected phases of the cycle shown in Fig. 5.2. $Da = 10^{-4}$, $Re = 500$, and $Wo = 11.3$

5.3.6 Pulsatile Flow in a Patient-Specific Geometry

Recent progress in medical imaging technology and improvements in computer hardware have enabled CFD analysis to predict arterial hemodynamics with increasing accuracy and reliability. In the present study, a patient-specific geometry is reconstructed from Magnetic Resonance Angiography (MRA) images using vascular modeling tools. The patient-specific geometry is obtained as an open source mesh created at the CISTIB laboratory of Universitat Pompeu Fabra in Barcelona (Spain). Using open source imaging tools GIMIAS and VMTK, the computational geometry has been constructed as shown in Fig. 5.12. At the inlet

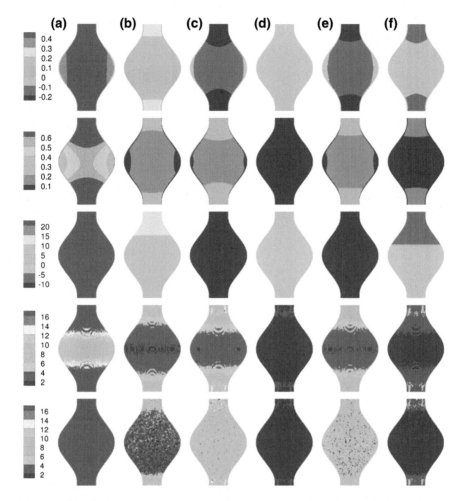

Fig. 5.11 Distribution of instantaneous axial velocity (at midplane), absolute velocity (at midplane), wall pressure, wall vorticity, and wall shear stress in porous medium at selected phases of the vascular cycle shown in Fig. 5.2. Da $= 10^{-4}$, Re $= 2000$, and Wo $= 11.3$

and outlet, a straight tube section is added such that boundary conditions are applied correctly and flow becomes practically fully developed before entering the patient-specific region. The inlet diameter is specified as 4.5 mm.

Assuming coil embolization as a treatment option for the cerebral aneurysm, simulations are discussed in the present section. Figure 5.13 shows points A–D where the flow variables, such as pressure and shear are probed before and after the coil embolization.

CFD analysis of blood flow in the above geometry (Fig. 5.12), with and without coil embolization, generates the pressure and shear waveform as shown in Fig. 5.14. For the purpose of simulation, the coil permeability is set to achieve a

Fig. 5.12 Open-source, patient-specific geometry for studying cerebral aneurysm

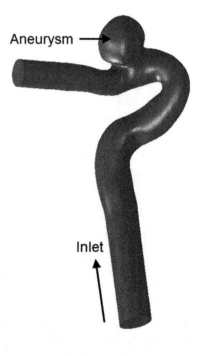

Fig. 5.13 Open-source, patient-specific geometry for studying cerebral aneurysm

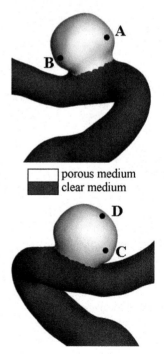

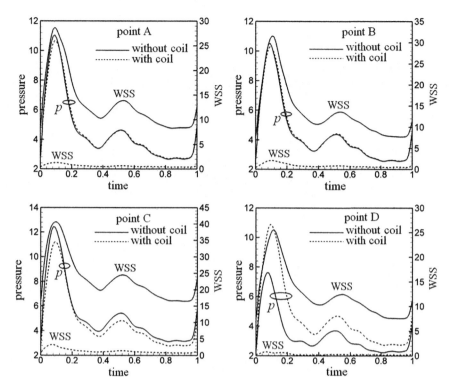

Fig. 5.14 Pressure and shear waveform at selected point of the aneurysm, before and after the coil embolization

Darcy number of 10^{-4}. As discussed in the previous section, at such a low Darcy number, the flow variables largely show diffusive behavior since viscous effects at the fluid–solid boundary are large. Figure 5.14 indicates that coil embolization greatly dampens the amplitude of shear stress, protecting the aneurysm from growth and rupture. The reduction in WSS is related to lowering of fluid velocity in the coil. In contrast, pressure drops are aggravated in the porous bulge. Accordingly, fluid pressure is found to be essentially independent of the coil-embolization treatment. Thus, pressure reduction in the bulge will require an independent medical procedure other than coil embolization.

5.4   Rheology of Biological Fluids

While existing models treat biological fluids as Newtonian, most biofluids, such as blood, often exhibit non-Newtonian rheology [60–65]. Blood approximately contains 55% plasma and 45% RBC. Other components in blood are present in very low quantities. The viscosity of blood is not constant and depends on several factors including plasma and its protein content, hematocrit, temperature, shear rate, and narrowness of the vessel. However, the effect of plasma content and temperature are much less compared to other factors. The concentration of plasma protein contributes to the blood viscosity. Blood viscosity increases with temperature at a very slow rate.

Hematocrit is defined as the percentage of blood volume made up of red blood cells. Hematocrit decreases from large vessel to small vessels. An increase in hematocrit results in rise of blood viscosity [66, 67].

For low shear rates ($\dot{\gamma} < 200$ s^{-1}), the variation of shear stress with strain rate is nonlinear. At higher shear rates, this relationship becomes linear. Generally, low shear rates are observed in small vessels (capillaries and small arteries), mainly due to lower flow rates and high shear rates are observed in large vessels. It follows that a Newtonian viscosity model may be adequate in large diameter arteries.

In addition to the effect of hematocrit and shear rate, small diameter of vessels (<500 μm) has another important phenomenon that influences blood viscosity; it is called the Fahraeus–Lindqvist effect. This effect considers the non-homogeneous nature of blood which is prominent when the corpuscles size becomes of the order of the vessel diameter. In this effect, the red cells move toward the center, leaving a layer of plasma at the wall [68].

5.4.1   Role of RBC in Blood Rheology

Plasma, in general, behaves as a Newtonian fluid. The presence of RBC contributes to the non-Newtonian behavior of blood. An RBC has 40% excess surface area compared to a sphere of the same volume [69]. This is attributed to the biconcave disc shape that t adopts under flowing conditions. The excess surface area allows the RBC to undergo area- and volume-preserving deformations, and increases the surface area available for gases to diffuse across the semi-permeable membrane [70].

The flow behavior of a suspension of RBCs in plasma resembles fluid droplets neutrally suspended in a fluid matrix. When blood is at rest, the RBCs aggregate into coin-stack-shaped structures called *rouleaux*. These structures break as the shear stress in blood increases, and RBCs become individually dispersed. At this stage, the dispersed RBCs preserve their biconcave shape and tumble in a flow with shear stress below 0.1 Pa [71]. The tumbling behavior gradually reduces, and the cells begin to align with the flow at around 0.2 Pa. The RBCs deform into an ellipsoidal shape, orient with the flow, and show tank-treading at shear stresses greater than ∼1 Pa. Tank-treading is a phenomenon in which the cell membrane rotates around the encapsulated fluid. As the shear rate increases, small pores open in the RBC membrane; the encapsulated hemoglobin oozes out of the blood cells into the bloodstream through these pores. This phenomenon is called hemolysis [72].

The aggregation of the RBCs into coin-stack structures, the breakage of these structures on imposed shear, and the deformation and orientation of the RBCs in flow make blood shear-thinning, i.e., the viscosity of blood decreases as the rate of strain in the flow increases. Moreover, viscoelasticity of the RBC membrane makes the whole blood viscoelastic as well. However, blood is weakly viscoelastic and its shear-thinning behavior is more prominent in small diameter vessels, for strain rates $<200 \text{ s}^{-1}$ [73].

If the viscosity of a fluid decreases with time, then the fluid is said to be *thixotropic*. At low shear rates, the microstructure formed by red blood cells is a three-dimensional structure. This structure requires a finite time to form and break leading to a dependence of viscosity and viscoelasticity on the duration of applied shear. This may contribute to the thixotropic behavior of blood. The difference between shear-thinning and thixotropy is as follows. In shear-thinning flow, viscosity will decrease as the shear rate increases; whereas in thixotropy, the viscosity at a given flow condition can decrease over time with same shear stress.

5.4.2 Realistic Blood Model

A realistic model of blood should consider the prominent characteristics of blood, in particular shear-thinning viscosity, viscoelasticity, and thixotropy. Since these characteristics are shear-dependent, a detailed picture of the flow field is first required. When blood flows through the vessels, the interaction between blood and solid walls (fluid–structure interaction—FSI) significantly affects hemodynamics. Therefore, FSI also needs to be incorporated in the mathematical model to obtain a realistic view of the cardiovascular system. However, blood is weakly viscoelastic and shear-thinning behavior is the more prominent non-Newtonian characteristic that must be accounted for in modeling. While the present study uses only Newtonian blood model, extension of non-Newtonian model in porous media is now an active area of ongoing research. Figure 5.15 provides a comprehensive summary for choosing a realistic blood model.

The above understanding of realistic blood rheology leads to three different modeling approaches in hemodynamic simulations. First is the Eulerian–Lagrangian approach where the plasma is treated as a continuum fluid and the RBCs are modeled as discrete, deformable particles. The second approach involves multiphase, two-fluid model where both plasma and RBCs are treated as continuous fluid phases interspersed with each other. In the third approach, blood is treated as a single-phase non-Newtonian fluid with shear-thinning behavior. A review of the literature shows two major approaches available for the single-phase, non-Newtonian blood model, namely the Carreau–Yasuda and the Casson models. These are given in terms of a flow-dependent viscosity as follows:

$$\text{Carreau--Yasuda:} \mu = \mu_\infty + (\mu_0 - \mu_\infty) + [1 + (\lambda \dot{\gamma})^a]^{(n-1)/a} \qquad (5.32)$$

$$\text{Casson:} \mu = \dot{\gamma}^{-1} \left[k_0 + k_1 \sqrt{\dot{\gamma}} \right]^2 \qquad (5.33)$$

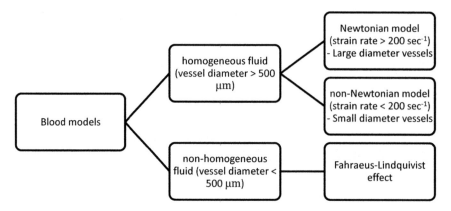

Fig. 5.15 Modeling criteria for realistic blood rheology

Here, a, n, λ, k_0, and k_1 are constants and the shear rate $\dot{\gamma}$ is defined as $\dot{\gamma} = 2\sqrt{D_{\mathrm{II}}}$ where D_{II} is the second invariant of the strain-rate tensor.

Both models show identical behavior at high shear rates (>1 s^{-1}). At a lower shear rate, the Carreau–Yasuda model provides a viscosity that asymptotically reaches a value of μ_0. The Casson model, however, shows an increasing trend of viscosity at lower shear rate leading to very large value as the shear rate approaches zero. Study shows that both the models are effective for hemodynamic simulations, while the Carreau–Yasuda model provides additional stability to numerical simulation [74, 75].

5.5 Closure

The present chapter describes modeling and simulation of porous media applications related to biological systems. Analytical solutions for flow in tubes filled with porous media are discussed. In particular, transport phenomena in an idealized aneurysm geometry filled with porous material is discussed. Results obtained from clear media and porous media are compared. It is evident that porous media modeling will play a significant role in understanding transport in the context of biofluid mechanics. While the present chapter deals with realistic pulsatile flow patterns in a reasonably complex geometry, patient-specific geometry should be considered for analysis. Further improvement in the rheological model of blood is also essential. Finally, modeling of non-Newtonian flow in porous media with continuously deformable vessel geometry and chemical reactions is a major challenge in simulation.

References

1. D.M. Sforza, C.M. Putman, J.R. Cebral, Hemodynamics of Cerebral Aneurysms. Annu. Rev. Fluid Mech. **41**, 91–107 (2009)
2. G.J. Sheard, Flow dynamics and wall shear-stress variation in a fusiform aneurism. J. Eng. Math. **64**, 379–390 (2009)
3. A. Quarteroni, S. Ragni, A. Veneziani, Coupling between lumped and distributed models for blood flow problems. Comput. Vis. Sci. **4**(2), 111–124 (2001)
4. S.C.M. Yu, J.B. Zhao, A steady flow analysis on the stented and non-syented sidewall aneurysm models. Med. Eng. Phys. **21**, 133–141 (1999)
5. R.B. Lieber, V. Livescu, L.N. Hopkins, A.K. Wakhiloo, Particle image velocimetry assesment of stent design influence on intra-aneurysmal flow. Ann. Biomed. Eng. **30**, 768–777 (2002)
6. T.M. Liou, S.M. Liou, K.L. Chu, Intra-aneurysmal flow with helix and mesh stent placement across side-wall aneurysm pore of a straight parent vessel. ASME J. Biomech. Eng. **126**, 36–43 (2004)
7. G. Canton, D.I. Levy, J.C. Lasheras, Changes in the intra-aneurysmal pressure due to hydrocoil embolization. AJNR Am. J. Neuroradiol. **26**, 904–907 (2005)
8. Y.P. Gobin, J.L. Counord, P. Flaud, J. Duffaux, In vitro study of hemodynamics in a giant saccular aneurysm model: influence of flow dynamics in the parent vessel and effects of coil embolization. Interventional Neuroradiology **36**, 530–536 (1994)
9. G. Canton, D.I. Levy, J.C. Lasheras, Hemodynamic changes due to stent placement in bifurcating intracranial aneurysms. J. Neurosurg. **103**, 146–155 (2005)
10. M.D. Ford, H.N. Nikolov, J.S. Milner, S.P. Lownie, E.M. DeMont, W. Kalata, F. Loth, D.W. Holdsworth, D.A. Steinman, PIV-measured versus CFD-predicted flow dynamics in anatomically realistic cerebral aneurysm models. ASME J. Biomech. Eng. **130**(021015), 021011–021019 (2008)
11. Q. Sun, A. Groth, M. Bertram, I. Waechter, T. Bruijns, R. Hermans, V.M. Pereira, O. Brina, T. Aach, Experimental validation and sensitivity analysis for CFD simulations of cerebral aneurysms. 2010 IEEE International Symposium of Biomedical Imaging: From Micro to Nano, pp. 1049–1052, 2010
12. L. Goubergrits, B. Thamsen, A. Berthe, J. Poethke, U. Kertzscher, K. Affeld, C. Petz, H.-C. Hege, H. Hoch, A. Spuler, In vitro study of near-wall flow in a cerebral aneurysm model with and without coils. Am. J. Neuroradiology **31**, 1521–1528 (2010)
13. M. Aenis, A.P. Stancampiano, A.K. Wakhloo, B.B. Lieber, Modeling of flow in a straight stented and nonstented sidewall aneurysm model. ASME J. Biomech. Eng. **119**, 206–212 (1997)
14. M. Ohta, S.G. Wetzel, P. Dantan, C. Bachelet, K.O. Lovblad, H. Yilmaz, P. Flaud, D.A. Rufenacht, Rheological changes after stenting of a cerebral aneurysms: a finite element modeling apporach. Cardiovasc. Intervent. Radiol. **28**, 768–772 (2005)
15. G.R. Stuhne, D.A. Steinman, Finite-element modeling of the hemodynamics of stented aneyrysm. ASME J. Biomech. Eng. **126**, 382–387 (2004)
16. D.A. Steinman, J.S. Milner, C.J. Norley, S.P. Lownie, D.W. Holdsworth, Image-basedcomputational simulation of flow dynamics in a giant intracranial aneurysm. Am. J. Neuroradiology **24**, 559–566 (2003)
17. V.L. Rayz, L. Boussel, G. Acevedo-Bolton, A.G. Martin, W.L. Young, M.T. Lawton, R. Higashida, D. Saloner, Numerical simulations of flow in cerebral aneurysms: comparison of CFD results and in vivo MRI measurements. J. Biomech. Eng. **130**, 1–9 (2008)
18. V.L. Rayz, L. Boussel, M.T. Lawton, G. Acevedo-Bolton, L. Ge, W.L. Young, R.T. Higashida, D. Saloner, Numerical modeling of the flow in intracranial aneurysms: prediction of regions prone to thrombus formation. Ann. Biomed. Eng. **36**, 1793–1804 (2008)
19. K. Takizawa, C. Moorman, S. Wright, J. Christopher, T.E. Tezduyar, Wall shear stress calculations in space–time finite element computation of arterial fluid–structure interactions. Comput. Mech. **46**, 31–41 (2010)

20. K. Takizawa, C. Moorman, S.I. Wright, J. Purdue, T. McPhail, P.R. Chen, J. Warren, T.E. Tezduyar, Patient-specific arterial fluid–structure interaction modeling of cerebral aneurysms. Int. J. Numer. Meth. Fluids **65**, 308–323 (2011)

21. A. Valencia, H. Morales, R. Rivera, E. Bravo, M. Galvez, Blood flow dynamics in patient-specific cerebral aneurysm models: the relationship between wall shear stress and aneurysm area index. Med. Eng. Phys. **30**, 329–340 (2008)

22. H.S. Byun, K. Rhee, CFD modeling of blood flow following coil embolization of aneurysms. Med. Eng. Phys. **26**, 755–761 (2004)

23. C. Groden, J. Laudan, S. Gatchell, H. Zeumer, Three-dimensional pulsatile flow simulation before and after endovascular coil embolization of a terminal cerebral aneurysm. J. Cereb. Blood Flow Metab. **21**, 1464–1471 (2001)

24. K.S. Cha, E. Balaras, B.B. Liebre, C. Sadasivan, A.K. Wakhloo, Modeling the interaction of coils with the local blood flow after coil embolization of intracranial aneurysms. ASME J. Biomech. Eng. **129**, 873–879 (2007)

25. K. Khanafer, R. Berguer, M. Schlicht, J.L. Bull, Numerical modeling of coil compaction in the treatment of cerebral aneurysm using porous medium theory. J. Porous Media **18**, 869–886 (2009)

26. T. Otani, M. Nakamura, T. Fujinaka, M. Hirata, J. Kuroda, K. Shibano, S. Wada, Computational fluid dynamics of blood flow in coil-embolized aneurysms: effect of packing density on flow stagnation in an idealized geometry. Med. Biol. Eng. Computing **51**(8), 901–910 (2013)

27. A.P. Mitsos, N.M.P. Kakalis, Y.P. Ventikos, J.V. Byrne, Haemodynamic simulation of aneurysm coiling in an anatomically accurate computational fluid dynamics model: technical note. Neuroradiology **50**, 341–347 (2008)

28. C. Paul, M.K. Das, K. Muralidhar, Three-dimensional simulation of pulsatile flow through a porous bulge. Transp. Porous Media **107**, 843–870 (2015)

29. V. Agarwal, C. Paul, M.K. Das, K. Muralidhar, Effect of coil embolization on blood flow through a saccular cerebral aneurysm. Sadhana **40**, 875–887 (2015)

30. J.R. Womersley, Method for the calculation of velocity, rate of flow and viscous drag in arteries when the pressure gradient is known. J. Physiol. **127**(3), 553–563 (1955)

31. I. Sazonov, S.Y. Yeo, R.L.T. Bevan, X. Xie, R.V. Loon, P. Nithiarasu, Modelling pipeline for subject-specific arterial blood flow - A review. Int. J. Numer. Methods Biomed. Eng. **27**, 1868–1910 (2011)

32. M. Abramowitz, I.A. Stegan, Handbook of mathematical functions, National Bureau of Standards Applied Mathematics Series, 1964

33. J.W. Peak, B.H. Kang, J.M. Hyun, Transient cool-down of a porous medium in pulsating flow. Int. J. Heat Mass Transf. **42**, 3523–3527 (1999)

34. S.Y. Byun, S.T. Ro, J.Y. Shin, Y.S. Son, D.Y. Lee, Transient thermal behavior of porous media under oscillating flow condition. Int. J. Heat Mass Transf. **49**, 5081–5085 (2006)

35. H. Dhahri, A. Boughamoura, S.B. Nasrallah, Forced pulsating flow and heat transfer in a tube partially filled with a porous medium. J. Porous Media **9**(1), 1–14 (2006)

36. H. Dhahri, A. Boughamoura, S.B. Nasrallah, Numerical study of heat transfer in a porous pipe subjected to reciprocating flow. J. Porous Media **9**(4), 289–305 (2006)

37. H. Dhahri, A. Boughamoura, S.B. Nasrallah, Entropy generation for pulsating flow in a composite fluid/porous system. J. Porous Media **11**(6), 557–574 (2008)

38. H. Dhahri, K. Slimi, S.B. Nasrallah, Viscous dissipation effects on heat transfer for oscillating flow in a pipe partially filled with a porous medium. J. Porous Media **2**(4), 381–395 (2010)

39. H.L. Fu, K.C. Leong, X.Y. Huang, C.Y. Liu, An experimental study of heat transfer of a porous channel subjected to oscillating flow. ASME J. Heat Transfer **123**, 162–170 (2001)

40. Z. Guo, S.Y. Kim, H.J. Sung, Pulsating flow and heat transfer in a pipe partially filled with a porous medium. Int. J. Heat Mass Transf. **40**(17), 4209–4218 (1997)

41. K. Habibi, A. Mosahebi, H. Shokouhmand, Heat transfer characteristics of reciprocating flows in channels partially filled with porous medium. Transp. Porous Media **89**, 139–153 (2011)

42. S.Y. Kim, B.H. Kangs, J.M. Hyun, Heat transfer from pulsating flow in a channel filled with porous media. Int. J. Heat Mass Transf. **31**(14), 2025–2033 (1994)
43. A.V. Kuznetsov, D.A. Nield, Forced convection with laminar pulsating flow in a saturated porous channel or tube. Transp. Porous Media **65**, 505–523 (2006)
44. A.V. Kuznetsov, D.A. Nield, Forced convection with laminar pulsating counterflow in a saturated porous circular tube. Transp. Porous Media **77**, 447–462 (2009)
45. K.C. Leong, L.W. Jin, An experimental study of heat transfer in oscillating flow through a channel filled with an aluminum foam. Int. J. Heat Mass Transf. **48**, 243–253 (2005)
46. T.V. Morosuk, Entropy generation in conduits filled with porous medium totally and partially. Int. J. Heat Mass Transf. **48**, 2548–2560 (2005)
47. A.A. Raptis, C.P. Perdikis, Oscillatory flow through a porous medium by the presence of free convective flow. Int. J. Eng. Sci. **23**(1), 51–55 (1985)
48. S. Chakravarty, A. Datta, Pulsatile blood flow in a porous stenotic artery. Math. Comput. Modell. **16**(2), 35–54 (1992)
49. M. El-Shahed, Pulsatile flow of blood through a stenosed porous medium under periodic body acceleration. J. Appl. Math. Comput. **138**(2–3), 479–488 (2003)
50. T. Gohil, R. McGregor, D. Szczerba, K. Burckhardt, K. Muralidhar, G. Szekely, Simulation of oscillatory flow in an aortic bifurcation using FVM and FEM: a comparative study. Int. J. Numer. Methods Fluids **66**, 1037–1067 (2012)
51. A.R. Khaled, K. Vafai, The role of porous media in modeling flow and heat transfer in biological tissues. Int. J. Heat Mass Transf. **46**, 4989–5003 (2003)
52. K.M. Khanafer, P. Gadhoke, R. Berguer, L.B. Joseph, Modeling pulsatile flow in aortic aneurysms: effect of non-Newtonian properties of blood. Biorheology **43**, 661–679 (2006)
53. K. Vafai, S. Kim, Forced convection in a channel filled with porous medium: an exact solution. ASME J. Heat Transfer **111**(4), 1103–1106 (1989)
54. K. Vafai, M. Sozen, Analysis of energy and momentum transfport for fluid flow through a porous bed. ASME J. Heat Transfer **112**, 690–699 (1990)
55. K. Vafai, K. and C. L. Tien, Boundary and inertia effects on flow and heat transfer in porous media. Int. J. Heat Mass Transf. **24**, 195–203 (1980)
56. K. Vafai (ed.), *Handbook of porous media* (Marcel Dekker, New York, 2005)
57. K. Vafai (ed.), *Porous media: Application in Biological Systems and Biotechnology* (CRC Press, Taylor and Francis group, 2011)
58. D.A. Nield, A. Bejan, Convection in Porous Media, Springer, New York, 2006
59. A.W. Date, Solution of transport equations on unstructured meshes with cell-centered colocated variables. Part I: discretization. Int. J. Heat Mass Transf. **48**, 1117–1127 (2005)
60. A. Narasimhan, The role of porous medium modeling in biothermofluids. J Indian Inst. Sci. **91**(3) (2011)
61. D. Arora, Computational Hemodynamics: Hemolysis and Viscoelasticity. Mechanical Engineering. University of Houston, Ph.D. thesis, 2005
62. T. Bodnár, A. Sequeira, M. Prosi, On the shear-thinning and viscoelastic effects of blood flow under various flow rates. Appl. Math. Comput. **217**, 5055–5067 (2011)
63. J. Janela, A. Moura, A. Sequeira, Absorbing boundary conditions for a 3D non-Newtonian fluid–structure interaction model for blood flow in arteries. Int. J. Eng. Sci. **48**, 1332–1349 (2010)
64. V. Kanyanta, A. Ivankovic, A. Karac, Validation of a fluid–structure interaction numerical model for predicting flow transients in arteries. J. Biomech. **42**, 1705–1712 (2009)
65. G.B. Thurston, Viscoelasticity of human blood. Biophys. J. **12**, 1205–1217 (1972)
66. M. Anand, K.R. Rajagopal, A shear-thinning viscoelastic fluid model for describing the flow of blood. Int. J. Cardiovasc. Med. Sci. **4**, 59–68 (2004)
67. K. Laganà, G. Dubini, F. Migliavacca, R. Pietrabissa, G. Pennati, A. Veneziani, A. Quarteroni, Multiscale modelling as a tool to prescribe realistic boundary conditions for the study of surgical procedures. Biorheology **39**, 359–364 (2002)
68. A. Leuprecht, K. Perktold, Computer simulation of non-Newtonian effects of blood flow in large arteries. Comput. Methods Biomechanics Biomechanical Eng. **4**, 149–163 (2001)

69. R.K. Dash, K.N. Meheta, G. Jayaraman, Casson fluid flow in a pipe filled with homogenious porous medium. Int. J. Eng. Sci. **34**(10), 1145–1156 (1996)
70. O.U. Mehmood, N. Mustapha, S. Shafie, Unsteady two-dimensional blood flow in porous artery with multi-irregular stenoses. Transp. Porous Media **92**, 259–275 (2012)
71. X. Zhang, Z. Yau, Y. Zhang, S. Xu, Experimental and computational studies on the flow field in aortic aneurysms associated with deployment of AAA stent-graft. Acta. Mech. Sin. **23**, 495–501 (2007)
72. A. Ogulu, E. Amos, Modeling pulsatile blood flow within a homogeneous porous bed in the presence of a uniform magnetic field and time-dependent suction. Int. Commun. Heat Mass Transfer **34**(8), 989–995 (2007)
73. S. Ramalho, A. Moura, A.M. Gambaruto, A. Sequeira, Sensitivity to outflow boundary conditions and level of geometry description for a cerebral aneurysm. Int. J. Numer. Methods Biomed. Eng. **28**, 697–713 (2012)
74. J. Boyd, J.M. Buick, S. Green, analysis of the Casson and Carreau-Yasuda non-Newtonian blood models in steady and oscillatory flows using the Lattice Boltzmann method. Phys. Fluids **19**, 0931031-14 (2007)
75. A.L. Fogelson, K.B. Neeves, Fluid mechanics of blood clot formation. Annu. Rev. Fluid Mech. **47**, 377–403 (2015)

Chapter 6
Oscillatory Flow in a Mesh-Type Regenerator

K. Muralidhar

Nomenclature

A_{IF} Specific area of the porous insert, $4(1 - \varepsilon)/d_w (\mathrm{m}^{-1})$
A_F Non-dimensional value of A_{IF} namely $A_{IF}R$
Bi Biot number for heat loss to the ambient hR/k_s
d_h Hydraulic diameter of the porous insert $\varepsilon d_w/(1 - \varepsilon)$ (m)
d_w Mesh wire diameter (m)
Da Darcy number, K/R^2
E Regenerator effectiveness, Eq. (6.44)
F_ε Ergun coefficient
f_j jth harmonic of friction factor equal to $|P_j|$
f_{eq} Total equivalent friction factor, Eq. (6.39)
h Heat transfer coefficient between the regenerator tube and the ambient (W/m²-K)
k Thermal conductivity (W/m-K)
K Permeability (m²)
Nu Fluid-to-solid Nusselt number at the scale of the mesh wire
p Non-dimensional fluid pressure scaled with ρW^2
P_j Pressure gradient for the jth harmonic scaled by $\rho W^2/R$
Pr Prandtl number
r Non-dimensional radial coordinate scaled with R
R Tube radius (m)
t Non-dimensional time scaled with R/W
T Temperature (K)
w Non-dimensional axial velocity scaled with W
W Velocity amplitude of pulsing flow (m/s)
$\overline{w}$ Cross-sectional mean velocity scaled with W
z Non-dimensional axial distance scaled with R
α_f Thermal diffusivity of the fluid (m²/s)
β Fluid-to-solid specific heat ratio
Γ Fluid-to-solid effective thermal conductivity ratio

© Springer International Publishing AG 2018
M.K. Das et al., *Modeling Transport Phenomena in Porous Media with Applications*, Mechanical Engineering Series,
https://doi.org/ 10.1007/978-3-319-69866-3_6

ΔT Reference temperature difference $T_H - T_L$ (K)
λ Fluid-to-solid thermal conductivity ratio
ε Porosity
φ_j Velocity amplitude of jth harmonic scaled by W
μ Dynamic viscosity of the working fluid (Pa-s)
ρ Fluid density (kg/m^3)
θ Non-dimensional temperature $(T - T_L)/(T_H - T_L)$
τ_c Thermal time constant of the porous insert (s)
ω Pulsing frequency of the flow scaled with ω_c
ω_c Characteristic frequency $\pi/\tau_c(\text{s}^{-1})$

Suffixes

f Fluid phase
H Hot half of the oscillation cycle
j jth harmonic
L Cold half of the oscillation cycle
R Quantity scaled with R as the length scale
w Quantity scaled with d_w as the length scale
s Solid phase

6.1   Introduction

In several thermal processes, hot fluids may be available from a part of the device or during a part of the cycle, before being exhausted to the ambient. The sensible thermal energy that would otherwise go waste can be used to heat up cooler fluids at some other part of the device or any other stage of the cycle. This transfer can be effected by introducing a regenerative heat exchanger (*regenerator*, in short) joining the two parts of the device or the two phases of the cycle. The regenerator stores and releases thermal energy and, thus, functions as a thermal sponge. In the present chapter, mathematical modeling of a regenerator in a Stirling refrigerator is discussed as an example and application of flow and transport in porous media.

6.1.1   Stirling Refrigerator

Stirling cycles are increasingly being used in specialized applications related to cooling, heating, and power generation. They use a non-condensing gas as a working medium whose thermodynamic changes occur entirely in a single phase. Stirling devices are, thus, structurally simple and easier to maintain and have a high reliability. Rapid advances are presently seen in Stirling cycles applied to cooling devices (called Stirling *cryocoolers*) used in space electronics and night vision imaging systems [1–5]. The term cryocooler is employed when the minimum temperature in the cycle falls below the liquid nitrogen temperature (<80 K).

The Stirling cycle is quite similar to the Joule–Brayton cycle that is employed in gas turbines, except that the regenerator is a necessary part in the former. Regenerators are also used in gas turbine cycles but to extract waste heat from exhaust gases in order to improve the cycle efficiency. In contrast, the regenerator enables the Stirling cycle to be completed and forms a primary component of the system.

For definiteness, consider a Stirling cycle running as a refrigerator (Fig. 6.1). Gaseous working fluid is filled in a part of a cylindrical tube bounded by the piston and the displacer at the two ends. In the middle, a regenerator is placed and two heat exchangers are mounted on each side. One of the heat exchangers transfers heat into the gas from a lower temperature heat source, namely the cold space. The second rejects heat from the gas to a high-temperature heat source, usually the ambient. The piston demarcates the point on the system boundary where work is imparted to the gas; in practice, the displacer can be replaced by a gas column piston.

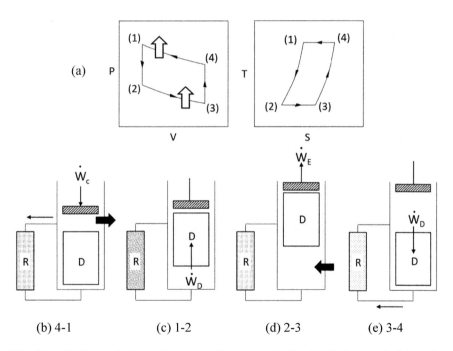

Fig. 6.1 a Stirling cycle functioning as a refrigerator on *P–V* and *T–S* diagrams. **b** Schematic drawing of a Stirling refrigerator with various stages of the cycle shown. Regenerator and displacer, respectively, are denoted as R and D. Symbol W is the work of compression **c** or expansion. **e**. Large arrows indicate heat transfer from and to the environment. Smaller arrows show the direction of gas flow

The regenerator of a Stirling refrigerator acts as a thermal sponge, absorbing energy into its solid phase when exposed to the hot gas and releasing it to the cold gas at a later stage in the cycle. Inside the regenerator, temperature distributes almost steadily but non-uniformly from the higher side temperature to the lower side.

Referring to Fig. 6.1 (left to right), the following observations may be noted. At a phase when the gas is being compressed by the piston, gas moves toward the displacer. The gas temperature is then raised so that heat is transferred from the gas to the higher temperature heat source through the high-temperature-side heat exchanger. Here, heat transfer is shown by a large bold arrow pointing right $(= Q_{out})$. For refrigerators, the high-temperature heat source is the ambient. The gas is still hotter within the regenerator than its time-averaged temperature, and the gas transfers thermal energy to the regenerator material. At the last stage of this phase, however, the displacer starts to move and the gas expands and is consequently cooled. Following this phase, the piston starts to move in the opposite direction and the gas is further cooled. Thus, the gas temperature becomes low enough to accept heat transfer from the lower temperature heat source, resulting in its own increase in temperature. The heat absorbed by the refrigerator from the cold source $(= Q_{in})$ is shown by a large bold arrow pointing leftward in Fig. 6.1. The refrigeration action is primarily associated with this part of the cycle. At the end of this phase, the gas temperature is still lower than the low-temperature side of the regenerator. Therefore, as the gas flows through the regenerator back toward the piston, the gas temperature increases. After a certain phase lag, the displacer reverses its motion, the gas is compressed, and its temperature becomes a little higher, equal to that of the stage one. The work input to the cycle is associated with the movement of the piston and the displacer. With reference to Fig. 6.1, it can be calculated as $(W_C + W_D - W_E)$. Hence, the device effectiveness defined in terms of the coefficient of performance (COP) is obtained as

$$COP = \frac{Q_{in}}{W_C + W_D - W_E}$$

The Stirling cycle can be represented on a P–V diagram as comprising two isothermal processes (for compression and expansion) and two constant volume processes (during the passage of the fluid in the regenerator), as in Fig. 6.1. When analyzed using classical tools of equilibrium thermodynamics, the cycle on the P–V diagram predicts a device performance that is identical to the Carnot cycle. This result is clearly unacceptable since experiments show that real Stirling systems have characteristics (such as COP) that are vastly inferior to the Carnot cycle predictions. Models that circumvent this difficulty by allowing non-ideal behavior, called *losses*, to be systematically included are needed. One such model for regenerators is discussed in the present chapter.

6.1.2 Regenerator

The Stirling refrigerator (cryocooler) sustains low temperatures in the cold space by expanding high-pressure low-temperature gas first available at the exit of the regenerator (point 2 on the *P–V* and *T–S* diagrams of Fig. 6.1). Since this process cannot be sustained indefinitely, low-temperature low-pressure gas available at the end of the cooling process is pushed into the regenerator (point 3, Fig. 6.1). The cyclic heating and cooling of the regenerator creates a dynamically steady temperature distribution in the regenerator. Thermodynamic analysis shows this variation to be linear, changing from the cold space temperature to the ambient [1, 4, 5]. The corresponding cycle COP can be shown to match the Carnot limit.

In the absence of a regenerator, a closed cycle device cannot be constructed and open systems such as gas turbines alone can be sustained. In this respect, a regenerator becomes an essential component of a Stirling refrigeration system.

An ideal regenerator will produce a linear temperature profile along its length under oscillatory conditions but zero pressure drop along its length. These idealized limiting conditions are possible only when a variety of conflicting requirements are fulfilled; these are listed below.

i. For stable operation, temperature fluctuations in the regenerator should be absent. Accordingly, an ideal regenerator must have a high heat capacity (thermal mass) and must be physically bulky.
ii. It must offer negligible resistance to flow, making it highly porous.
iii. The dead space in the regenerator, i.e., the space occupied by the stagnant fluid, must be small indicating that the porosity be low.
iv. The heat transfer rate between the solid material and the flowing fluid must be large. Consequently, the regenerator matrix must be finely divided to create a large surface area.
v. The regenerator must have low thermal conductivity in the flow direction to prevent temperature equalization and ensure a linear temperature variation required from a thermodynamic viewpoint. It should be a large value transverse to it, to ensure uniformity of temperature in a plane normal to the flow direction.

It is clear that the foregoing conditions cannot be simultaneously satisfied. As a result, real-life regenerators display a performance that is inferior to the ideal. The corresponding refrigerator performance is also suboptimal, falling below that of the Carnot cycle. Practical regenerators in cooling applications either comprise fine bronze meshes stacked next to each other or lead shots forming a porous bed. The latter is particularly preferred at low temperatures (in the 1–5 K range) where the thermal capacity of most metals (except lead) becomes quite small. At still lower temperatures, regenerators have to be substituted by recuperators that resemble a standard counterflow heat exchanger [6].

Most Stirling cycle models in the published literature treat the regenerator as *perfect*. In quantitative terms, it requires the temperature variation to be linear along the regenerator while temperature fluctuations are strictly zero. The pressure drop (measured in terms of an RMS value) is also neglected. These assumptions are occasionally relaxed by introducing a heat transfer correlation between the fluid and the solid phases as well as one for pressure drop. For a realistic full cycle simulation of the device, it is important to determine pressure drop and temperature distribution under pulsatile flow conditions. Related quantities of importance are the time-averaged and RMS solid and fluid phase temperatures, their amplitudes, and temporal relationships, along with the associated time constants. Determination of these quantities over a wide range of conditions of geometry, frequency, flow rates, and temperatures is of practical importance. While the governing equations appropriate for Newtonian fluids (with corrections for variation in fluid properties) can be used in the simulation, the geometry of the flow space within the regenerator is quite complex. The present chapter adopts a porous medium approximation of the regenerator to analyze flow and heat transfer.

6.2 Thermodynamic and Transport Models

Two families of mathematical models that can be used for Stirling cycle analysis are to be distinguished: *thermodynamic* and *transport*. In the former, rate processes that drive flow and heat transfer do not appear in the formulation. The working fluid undergoes a series of states (in pressure and temperature) forming a cycle. The cyclic process can be sustained only when a certain amount of net work is transferred to the fluid, leading to a net transfer of heat from a low-temperature reservoir to one at a higher temperature. The thermodynamic model is to an extent simple in the sense that spatial variation in the flow properties such as velocity, pressure, and temperature are approximated (idealized) or completely neglected. A majority of thermodynamic models of Stirling cryocoolers published till date treat the fluid pressure to be spatially constant, though a function of time. Temperature, on the other hand, has a time-frozen spatial distribution, varying from the hot space to the cold, via the regenerator. The compression and expansion processes are treated as isothermal, though refined thermodynamic models at the level of a gas particle, adopt an approach often referred as *Lagrangian*. Such models have demonstrated successfully the origin of losses in real cryocoolers [7]. Thermodynamic models are also analytically tractable, to form a part of an overall optimization algorithm.

The transport model differs from the thermodynamic one in the sense that the flow quantities depend on both space and time. The formulation is set up from basic physical laws such as conservation of mass and energy, and Newton's second law of motion. In this approach, the frictional pressure drop, temperature variation (linear or otherwise), and heat exchanger effectiveness arise as a part of the solution. The flow and thermal fields continuously develop in time, till a steady state is reached. In general, a substantial number of details can be built into this model. A major weakness of this approach is the resulting mathematical complexity of the governing equations and concomitant cost of the computational effort. Very few transport models of cryocoolers are available in the literature; for a one-dimensional version, see Organ [5]. However, the approximation-free nature of the model makes it very attractive, and the need for such models has often been highlighted.

Examples of well-established thermodynamic models of Stirling devices (engines as well as cryocoolers) are the ideal cycle model and the Schmidt cycle model described in detail by Walker [1]. These models are essentially loss-free and predict COPs that are equal to the Carnot value. They have limited utility and can be used to examine the role of pressure, temperature, and RPM on quantities such as power input and heat absorbed.

In the present chapter, a detailed model that can predict regenerator losses is presented. The central idea is to represent a regenerator mesh as a homogeneous isotropic porous medium. The performance of a regenerator as a function of porosity, length, and the device frequency discussed here follows an earlier work of one of the authors [8]. The approach of studying transport phenomena in a mesh using porous medium modeling can also be seen in several other references [9–12].

6.3 Transport Modeling of a Mesh-Type Regenerator

An ideal regenerator is one that experiences zero pressure drop in both steady and oscillatory flows, while creating a linear temperature variation between the hot and the cold spaces of the device. The linearity is preserved even under oscillatory conditions of flow. In addition, temperature is spatially uniform in the direction transverse to the main flow. Clearly, the performance of a real-life regenerator will depart from the ideal. The extent of departure from ideality will depend on the mesh structure, flow conditions such as Reynolds number and pulsing frequency, and the thermal properties of the gas and the material of the solid phase. The prediction of velocity, pressure, and temperature, apart from global properties such as friction factor and Nusselt number, requires a transport model of mass, momentum, and energy transfer. These equations can be solved to determine the regenerator properties, expressed in terms of the friction factor, effectiveness, linearity of the temperature profile, and the amplitude of temperature fluctuations. It is to be expected that the regenerator performance will be an optimum under certain conditions and will deteriorate otherwise. This chapter studies the performance of a regenerator mesh in a Stirling cycle via a porous medium model with an eye on enabling its design and selection in practical problems.

A regenerator mesh will resemble a classical porous medium (e.g., a bed of spheres) when considerably dense. A coarse mesh will appear as a collection of cylindrical rods. At these extremes, classical limits of Darcian flow (dense mesh) and flow past a circular cylinder (coarse mesh) will hold. These limits are contained in the non-Darcy model presented in Chap. 2, wherein parameters such as permeability and the inertia coefficient are to be suitably prescribed. In fact, the non-Darcy model includes the pure fluid limit of porosity approaching unity and the permeability approaching infinity. Thus, the non-Darcy flow model is expected to be appropriate for regenerator modeling.

Two factors which absorb the effect of geometry of the pore space of the mesh in the model are porosity and permeability. For porous regions that are formed as a collection of cylinders or spheres of uniform diameter, porosity and permeability are interrelated and one can be derived from the expression for the other. Exceptions are permeability–porosity relations near a solid wall (even for an assembly of spheres and cylinders) where data must come from experiments. When the porous region cannot be thought of as an assembly of uniform diameter cylinders (or spheres), the porosity–permeability relation breaks down and permeability must be determined from steady flow experiments.

Non-Darcy models of porous media that account for the presence of a wall, inertia, and viscous effects in the fluid phase and vortex formation in the pores have been widely reported [13–18]. Using this approach, Vafai and Kim [17] developed an exact solution for one-dimensional flow and heat transfer in fully developed flow in a porous channel with heated walls. The authors obtained conditions under which both wall and inertia effects are significant. Lage [19] derived order-of-magnitude estimates of transient wall heat transfer rates in a rectangular porous enclosure using scale analysis. The expressions for Nusselt number explicitly bring out the contributions of gravity, inertial, viscous, and transient terms in the governing equations. Antohe and Lage [20] studied buoyancy-driven time-periodic flows in a rectangular enclosure heated from below. As the medium became increasingly porous, resonance was seen at selected Rayleigh numbers. The peak heat transfer rates reached a maximum at the pure fluid limit.

6.3.1 Thermal Non-Equilibrium

One of the assumptions often adopted in porous media models is thermal equilibrium between the fluid and the solid phases locally within the representative elementary volume [21, 22]. Such an assumption is adequate at steady state or during transients that are slow enough with respect to both the solid- and the gas-side time constants. In all other cases, the two phases must be assigned individual temperatures along with the relation connecting them at their interface. Loss of thermal equilibrium is of primary importance in the study of regenerators whose performance requires energy exchange between the solid and the fluid phases with widely different thermal properties.

Sozen and Kuzay [23] studied enhanced heat transfer in a heated tube using mesh screens using the approximation of thermal equilibrium. The permeability of the screens was determined from experiments. Kim et al. [24] studied heat transfer in a channel filled with porous material when subjected to oscillatory flow superimposed on a nonzero mean. While the Darcy equations were extended to include inertial and viscous effects, the authors continued to assume thermal equilibrium.

Kuznetsov [25–27] analytically investigated the effect of non-equilibrium between the solid and fluid phase temperatures on heat transfer in packed beds and semi-infinite media. Results were obtained under simplifying conditions of uniform (constant) velocity and simple geometry and with an experimentally determined heat transfer coefficient between the solid and fluid surfaces. Observations arising from these studies are a phase difference between the waves propagating through each of the phases and the temperature difference wave localizing in space. Kuznetsov and Vafai [28] analyzed flow under thermal nonequilibrium including reactions in a metal hydride bed. The authors determined conditions under which approximations such as thermal equilibrium, filtration, and the frontal model are valid. In a related work, Vafai et al. [29–31] numerically studied single- as well as two-phase flow of a fluid (and associated heat transfer) in a porous medium. The roles of inertial and viscous forces, development length, effect of porosity, and proximity to a wall were investigated under thermal non-equilibrium conditions. Their results showed that lack of thermal equilibrium was most significant at high Reynolds numbers, for example, at high velocities and in gases and high Darcy numbers (i.e., high porosity systems). These results are of importance in modeling regenerators where one encounters large gas velocities and high porosity mesh screens [32–34].

6.4 Non-Darcy Thermal Nonequilibrium Model

In the following discussion, flow and heat transfer in a regenerator mesh are taken to be analogous to transport processes that occur in a fluid-filled porous medium. Features specific to a regenerator are included in the analysis. Specifically, inertial effects in the fluid phase are retained since the working fluid is a gas and Reynolds numbers are quite large. Viscous terms are required to model wall interactions and boundary-layer formation. Suitable porosity and permeability functions for the mesh and experimentally determined friction factor relationship are utilized to establish the correct momentum equation for the porous continuum. Thermal non-equilibrium between the gas and the metallic wire of the mesh is allowed by specifying a Nusselt number relation linking the gas and the metallic wire temperatures. A model thus developed is driven by oscillatory flow, with hot and cold fluid alternately moving past the mesh.

The regenerator mesh is an insert in a tube of radius R and length L as shown in Fig. 6.2. The mass, momentum, and energy equations for flow and heat transfer in a porous media under thermal nonequilibrium conditions are obtained from Chap. 2. These equations as applicable for a regenerator mesh are presented below in dimensionless coordinate-free form (see Sect. 2.4, Chap. 2):

Fig. 6.2 Geometry of a regenerator mesh and coordinate system. Hot fluid enters at a temperature T_H on the left side and cold fluid at T_L on the right

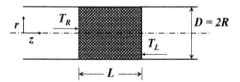

mass: $\nabla \cdot \mathbf{u} = 0$

momentum:

$$\frac{1}{\varepsilon}\frac{d\mathbf{u}}{dt} = \frac{1}{\varepsilon}\left(\frac{\partial \mathbf{u}}{\partial t} + \frac{1}{\varepsilon}\mathbf{u}\cdot\nabla\mathbf{u}\right) = -\nabla p - \frac{1}{Re\,Da}\mathbf{u} - \frac{F}{\sqrt{Da}}\mathbf{u}|\mathbf{u}| + \frac{1}{\varepsilon\,Re}\nabla^2\mathbf{u} \qquad (6.1)$$

Fluid:

$$\varepsilon\left(\frac{\partial T_f}{\partial t} + \frac{\mathbf{u}}{\varepsilon}\cdot\nabla T_f\right) = \frac{1}{Re\,Pr}\nabla\cdot\left(\frac{k_{\text{eff,f}}}{k_f}\right)\nabla T_f - \frac{Nu}{Re\,Pr}A_{\text{IF}}(T_f - T_s)$$

Solid:

$$(1 - \varepsilon)\frac{\partial T_s}{\partial t} = \frac{\beta}{Re\,Pr}\nabla\cdot\left(\frac{k_{\text{eff,s}}}{k_f}\right)\nabla T_s + \frac{Nu}{Re\,Pr}A_{\text{IF}}\beta(T_f - T_s)$$

$$(6.2)$$

Here, Darcian velocity $\mathbf{u}$ (the REV-averaged value) is scaled by the amplitude of velocity oscillation U, distances and coordinates by the tube radius, temperature by the difference between the hot and the cold fluid temperatures, and pressure by ρU^2. The effective thermal conductivity (shown by the suffix 'eff') includes dispersion in the fluid phase in the fluid temperature equation; the corresponding contribution in the solid phase is zero [35–37]. Suffix 'f' indicates fluid properties, while symbol 's' is for the solid phase. The definition of Prandtl number is based on the fluid properties. In subsequent discussions, the non-dimensional effective thermal conductivities are written in compact form as

$$\Gamma_f = \frac{k_{\text{eff,f}}}{k_f}; \quad \Gamma_s = \frac{k_{\text{eff,s}}}{k_f} = \frac{(1 - \varepsilon)k_s}{k_f} = \frac{1 - \varepsilon}{\lambda}$$

Effective thermal conductivity in the fluid phase can, in general, be a tensorial quantity. In the present discussion, effective thermal conductivity of the fluid is taken as purely diagonal with components Γ_r and Γ_z in the r- and the z-directions. In Eq. 6.2, β is the fluid-to-solid heat capacity ratio

$$\beta = \frac{(\rho C)_f}{(\rho C)_s}$$

The momentum equation is written for the Darcian velocity, fluid velocity in the pore space of the mesh being obtained as $\mathbf{u}/\varepsilon$. The fluid and solid phase temperatures are coupled by the interphase heat transfer rate prescribed in terms of Nu, the Nusselt number. The applicable reference area is the interfacial area per unit volume, scaled by the tube radius and denoted as A_{IF}. Dimensionless quantities Re, Pr, and Pe are defined as

$$Re = \frac{\rho UR}{\mu}; \quad Pr = \frac{(\mu/\rho)_f}{(k/\rho C)_f} = \left(\frac{\mu C}{k}\right)_f; \quad Pe = Re \times Pr = \frac{UR}{(k/\rho C)_f} \tag{6.3}$$

The interphase Nusselt number is defined in terms of the fluid-to-solid heat transfer coefficient as

$$Nu = \frac{hR}{\alpha_f}; \quad \alpha_f = \left(\frac{k}{\rho C}\right)_f$$

Other symbols are defined in the nomenclature at the end of this chapter. The inertia coefficient F (also called the Ergun factor), interphase Nusselt number Nu, and dispersion k_{eff} (in terms of Γ_r and Γ_z) are to be determined from dedicated, but simplified experiments, as outlined in Sect. 6.4.2.

6.4.1 Flow Equations in One-Dimensional Unsteady Form

While the above formulation is quite general, the following simplifying assumptions are adopted for the purpose of obtaining semi-analytical solutions of the flow field.

a. The regenerator mesh is taken to be placed in a tube as an insert, and calculations are carried out in the axisymmetric (r-z) coordinate system.
b. Flow is incompressible.
c. Flow is fully developed though a function of time. Hence, only the axial velocity component is nonzero and a function of the radial coordinate and time.
d. The oscillatory flow field is assumed to reach dynamic steady state much faster than the thermal field. Accordingly, the flow and energy equations are decoupled. With the flow field prescribed, the thermal fields in the two phases are then determined numerically from a system of equations written in axisymmetric coordinates.

Typical gas velocities that can be expected in a Stirling refrigerator are 5–10 m/s, being considerably smaller than the sonic velocity. The assumption of incompressibility is thus justified. Additionally, the pressure drop across the mesh is not expected to be large, though the fluid experiences large changes in pressure and density in the rest of the thermodynamic cycle. For the third assumption, it should be noted that regenerator meshes are sufficiently dense and the fully developed velocity profile is flat over a major portion of the tube cross section. As estimated in [15], wall effects are confined to a dimensionless distance of the order of $\sqrt{Da}$ ($R\sqrt{Da}$ in dimensional form). Since Da $\sim 10^{-6}$, wall effects are within 10^{-3} (0.1%) of the tube radius. Correspondingly, one can expect the development distance to be small and hence negligible. This assumption must, however, be dropped in high porosity meshes, particularly at high Reynolds numbers.

The justification for the fourth assumption (d) on transient durations is the following. Flow transients will last for an extent of time that depends on the ratio of the distance between solid surfaces and the average fluid velocity. For the mesh, this distance is the wire spacing, which is a small quantity, when compared to the tube radius. The implication is that the flow transient will continue till the boundary conditions between adjacent surfaces (the wires, for a mesh) are established. In contrast, the thermal transients are determined by the heating and cooling rates of the wire, and, in turn, the gas-to-solid heat transfer coefficient and the thermal capacity of the wire material. The ratio of the timescales of flow to heat transfer can be estimated as

$$\frac{\text{timescale of flow}}{\text{timescale of heat transfer}} = \frac{Nu \times \beta}{Re\,Pr}$$

Here, β is the ratio of the thermal capacity of the fluid (a gas such as air or helium) to the solid (a bronze alloy); see Sect. 6.4.3. This ratio is much smaller than unity, confirming that flow is established quickly while thermal transients are delayed.

With the assumptions stated earlier for a regenerator mesh, Eqs. 6.1–6.3 in the r-z coordinate system can be stated as follows. For fully developed flow, the transverse velocity components u and v are zero and the only nonzero velocity component is along the axial direction. The mass balance equation simplifies to the form:

$$\text{Mass:} \quad \frac{\partial w}{\partial z} = 0 \tag{6.4}$$

Since only one velocity component is nonzero (but oscillatory), the system of three momentum equations reduces to one for the axial velocity component and can be derived as:

Momentum:

$$\frac{1}{\varepsilon}\frac{\partial w}{\partial t} = -\frac{dp}{dz} - \frac{1}{Re\,Da}w + \frac{1}{\varepsilon Re}\left(\frac{\partial^2 w}{\partial r^2} + \frac{1}{r}\frac{\partial w}{\partial r}\right) - \frac{F}{\sqrt{Da}}w^2 \tag{6.5}$$

The thermal energy equations in the fluid and the solid phases simplify in the r-z coordinate system as follows:

Fluid temperature:

$$\varepsilon\frac{\partial \theta_f}{\partial t} + w\frac{\partial \theta_f}{\partial z} = \frac{1}{Re\,Pr}\left\{\frac{1}{r}\frac{\partial}{\partial r}\left(r\Gamma_r\frac{\partial \theta_f}{\partial r}\right) + \frac{\partial}{\partial z}\left(\Gamma_z\frac{\partial \theta_f}{\partial z}\right)\right\}$$
$$- \frac{NuA_{IF}}{Re\,Pr}(\theta_f - \theta_s) \tag{6.6}$$

Solid temperature:

$$(1 - \varepsilon) \frac{\partial \theta_s}{\partial t} = \frac{\beta}{\lambda} \frac{1}{\text{Re Pr}} (1 - \varepsilon) \left\{ \frac{\partial^2 \theta_s}{\partial r^2} + \frac{1}{r} \frac{\partial \theta_s}{\partial r} + \frac{\partial^2 \theta_s}{\partial z^2} \right\}$$
$$+ \frac{Nu A_{IF} \beta}{\text{Re Pr}} (\theta_f - \theta_s) \tag{6.7}$$

The symbols appearing in these equations are non-dimensional and carry the usual meaning. The scales employed for this purpose are selected from the regenerator application under discussion. Symbol w is the non-dimensional velocity component along the non-dimensional axial coordinate z normalized by the amplitude of velocity fluctuation W, r the non-dimensional radial coordinate, and θ the non-dimensional temperature which is the difference from the reference temperature T_C and normalized by the overall temperature difference $\Delta T = T_H - T_C$. Further, we have Nu the Nusselt number defined with R as the representative length scale, F the inertia coefficient, Pe the Peclet number, A_{IF} the non-dimensional specific surface area of porous insert, ε the porosity, β and λ the fluid-to-solid ratios of heat capacity and thermal conductivity, Γ_r and Γ_z the effective thermal conductivity of the fluid medium in the r- and the z-directions, and subscripts f and s are for fluid and solid phases, respectively.

The flow and heat transfer equations given above must be solved, along with suitable initial and boundary conditions. Initial conditions generally correspond to quiescent flow and thermal conditions; that is, all velocities are zero and temperatures are uniform (and equal to zero in a dimensionless context). When a harmonic analysis technique is used, an initial condition is not required for velocity (Sect. 6.4.4). Flow could be driven by a velocity prescribed at the inflow plane or a prescribed pressure difference, both of which will be sinusoidal functions of time. Since a Stirling cycle employs positive displacement devices such as a piston, it is more appropriate to prescribe the flow (and hence the cross-sectional averaged velocity). The temperature boundary condition is applied in the following manner. When the fluid enters the mesh on the left plane, it is assigned a temperature of unity. When the sign of the fluid velocity on the inflow plane reverses, the inflow plane for the temperature calculation is taken as the right-side boundary and the fluid is assigned a zero temperature. Thus, the inflow and outflow planes are periodically switched for the temperature calculation from one half-cycle to the next. The outflow condition for temperature is of the zero gradient type. The no-slip condition for velocity and an adiabatic condition for temperature are applicable at the tube wall. For regenerators that are long with respect to the tube radius, thermal losses to the ambient cannot be neglected. The temperature boundary condition for the fluid phase is then written as

$$-\varepsilon \frac{\partial \theta_f}{\partial r} \bigg|_{r=1} = \frac{Bi}{\lambda} \left(\theta_f \big|_{r=1} - \theta_\infty \right) \tag{6.8}$$

Similarly, for the solid phase:

$$-(1-\varepsilon)\frac{\partial \theta_s}{\partial r}\bigg|_{r=1} = Bi(\theta_s|_{r=1}-\theta_\infty) \tag{6.9}$$

Here, Bi is the Biot number, θ_∞, the ambient temperature set equal to $\theta_H\,(=1)$, and λ, the fluid-to-solid thermal conductivity ratio. In the present study, heat loss to the ambient is prescribed to be zero, i.e., $Bi = 0$ except for a limited number of cases where a value 3×10^{-3} is prescribed (Sect. 6.5.4). These exceptional studies are performed to see qualitatively the effect of heat exchange with the ambient on the regenerator performance. Computation is started with uniformly distributed non-dimensional temperature of zero. Since heat losses take place from the fluid as well as the solid portions of the mesh, the wall gradient boundary condition for heat loss is corrected by a factor ε for the fluid and a factor $(1-\varepsilon)$ for the solid. The geometry employed is axisymmetric, and along the axis $(r = 0)$, the symmetry condition

$$\begin{aligned} r=0: \quad & \frac{\partial w}{\partial r}=0 \\ r=0: \quad & \frac{\partial \theta_f}{\partial r}=\frac{\partial \theta_s}{\partial r}=0 \end{aligned} \tag{6.10}$$

is applied in all the calculations.

6.4.2   Specification of Model Parameters

Equations 6.5–6.10 need the specification of a several parameters before a solution can be developed. The working fluid is taken as air in the present chapter, with a Prandtl number of 0.7. A quantity such as porosity (ε) can be independently found by measuring the volume of liquid (say, water) required to fill a tube containing the porous region. The flow parameters appearing in the momentum equation are permeability K and the inertia coefficient F. These are constants for a given mesh and hence can be found from steady flow experiments. In a typical experiment, the apparatus is a tube carrying the mesh of given porosity and length L. The measured quantities are the average velocity W and the pressure drop Δp. The steady-state momentum equation with a uniform velocity distribution in the radial direction is derived from Eq. 6.5 as:

$$0 = -\frac{\Delta p}{L}-\frac{1}{Re\,Da}w-\frac{F}{\sqrt{Da}}w^2$$

It follows that experimental measurements can be correlated as

$$-\frac{\Delta p}{L} = AW + BW^2 \qquad (6.11)$$

Here, A and B are constants to be found from curve-fitting the quadratic form in velocity to the pressure measurements. For a uniform mesh of constant porosity, the velocity distribution is practically flat across the cross section and the wall shear stress is zero. Hence, the pressure drop has contributions mainly due to viscous friction within the mesh and inertia effects. Comparing the quadratic relation (6.11) to a spatially integrated form of the governing Eq. (6.5), permeability and porosity can be obtained. It can be shown that wall effects, if important, can be accounted for in the constant A. In this respect, neglecting wall shear stress is equivalent to under-predicting the mesh permeability. The interphase Nusselt number of the energy equation is the third parameter that must be experimentally determined.

Parameters of the momentum and energy equations have not been simultaneously prescribed for any single regenerator mesh in the literature. Chen et al. [38] report some of them for a particular mesh from experiments with air as the working fluid. Hence, the majority of simulation reported in this work pertains to this mesh. Sozen and Kuzay [23] report the hydrodynamic parameters for a mesh used in heat transfer augmentation. This mesh also has been analyzed in the present work for comparison with the mesh of [38].

Parameters of the mesh reported in [38] and of its equivalent porous insert placed in a tube are as follows:

porosity $\varepsilon = 0.703$, wire diameter $d_w = 0.0508$ mm, mesh hydraulic diameter (estimated from the geometry of an array of equi-spaced cylinders) $d_h = \frac{\varepsilon}{1-\varepsilon} d_w = 0.1202$ mm, tube radius $R = 9.5$ mm, pore surface area parameter (estimated from the geometry of an array of equi-spaced cylinders)

$$A_{IF} = \frac{4(1-\varepsilon)R}{d_w} = 222.16\,\text{m}^{-1}$$

The dimensionless relationship for the measured pressure drop under steady flow conditions is reported by the authors as follows [38]:

$$-\varepsilon^2 \left(\frac{1}{[\rho W^2/2]}\frac{\Delta p}{L/R}\right)\frac{d_h}{R} = \frac{174.1\varepsilon}{\text{Re}}\frac{R}{d_h} + 2.645; \quad \text{Re} = \frac{\rho WR}{\mu} \qquad (6.12)$$

Parameters K and F can be found by comparing Eq. 6.12 with Eqs. 6.5 and 6.11. The integrated form of Eq. 6.5 can be written in the following form:

$$-\varepsilon^2 \frac{dp}{dz}\frac{d_h}{R} = \frac{2\varepsilon^2}{\text{Re}Da}\frac{d_h}{R} + \frac{2F\varepsilon^2}{\sqrt{Da}}\frac{d_h}{R} \qquad (6.13)$$

Comparison of the above two relationships (Eqs. 6.12 and 6.13) yields the equivalent permeability $K = 1.172 \times 10^{-10} \mathrm{m}^2$ and the corresponding Darcy number

$$Da = \frac{K}{R^2} = 1.298 \times 10^{-6}.$$

Further, the inertia factor can be extracted by comparison with Eq. (6.13) as $F = 0.24$.

The interphase heat transfer coefficient is obtained from a transient experiment in which cold fluid passes through an initially heated mesh or hot fluid passes through an initially cold mesh. The mesh temperature is then recorded as a function of time. Writing this variation in terms of the dimensionless wire temperature θ as

$$\theta = C \exp(-Dt) \tag{6.14}$$

parameters C and D can be determined. These, in turn, depend on the flow rate and hence the Reynolds number. Within a narrow range of temperatures, fluid properties can be taken to be a constant. In general, however, heat transfer rates will depend on Reynolds as well as the Prandtl number. Since in the present discussion, $Pr = 0.7$, an explicit dependence on Pr is not shown. Parameter D is the reciprocal of the thermal time constant of the mesh and can be related to the interphase heat transfer coefficient (h) via a lumped parameter model. Hence

$$D(\mathrm{Re}) = \frac{hA_I}{(\rho C)_s} \tag{6.15}$$

Here, ρ is the density of the mesh material, C is its specific heat, and A_I is the (solid-fluid) interfacial area per unit volume $(= A_{IF} \times R)$. With the heat transfer coefficient determined, the interphase Nusselt number as a function of Reynolds number can be obtained.

The solid-to-fluid heat transfer data of Chen et al. [38] is expressed in the following form:

$$Nu_h = 0.483 \mathrm{Re}_h^{0.548}$$

Changing the length scale from the hydraulic diameter d_h to the tube radius, one obtains

$$Nu = 3.47 \, \mathrm{Re}^{0.548} \tag{6.16}$$

This correlation is used in the solution of the energy equations of the regenerator mesh.

The mesh parameters reported by Sozen and Kuzay [23] can be summarized as:

$\varepsilon = 0.85$, $d_w = 0.2$ mm, hydraulic diameter $d_h = 1.133$ mm, tube radius $R = 5$ mm.

As seen by its porosity, the mesh here is coarse and offers smaller resistance than the one of Chen et al. [38]. The equivalent values of permeability and the Darcy number are

$$K = 6.99 \times 10^{-9} \text{m}^2 \, Da = 3.0 \times 10^{-4}$$

The experimentally determined inertia coefficient of Sozen and Kuzay [23] is given as

$$F = 0.05$$

The solid-to-fluid Nusselt number relation is not specified in this study. Since the mesh porosity is quite high, each wire of the mesh can be visualized as a single cylinder and the following applicable correlations in air can instead be used [39, 40]:

$$\begin{aligned}
\text{Re}_w < 4 \, Nu_w &= 0.891 \text{Re}_w^{0.330}; \\
4 < \text{Re}_w < 40 \, Nu_w &= 0.821 \text{Re}_w^{0.385}; \\
40 < \text{Re}_w < 4000 \, Nu_w &= 0.615 \text{Re}_w^{0.466}
\end{aligned} \qquad (6.17)$$

Here, suffix w indicates quantities evaluated with the wire diameter as the length scale. Under oscillatory flow conditions, Reynolds number is based on the velocity amplitude W and not the instantaneous velocity. The expression can be suitably re-derived in terms of the tube radius.

Applying the expressions developed for the porous media composed of spherical solid bodies [21, 22, 35], effective thermal conductivity of the mesh normalized with the fluid phase thermal conductivity is given for air (Pr = 0.7) by the following relationships:

$$\begin{aligned}
\Gamma_r &= \varepsilon + 0.05 \text{Re}_w \, \text{Pr} \\
\Gamma_z &= \varepsilon + 0.1 \text{Re}_w \, \text{Pr}
\end{aligned} \qquad (6.18)$$

As before, Re_w is the Reynolds number based on the wire diameter and the velocity amplitude. This expression includes the effect of the thermal dispersion in the fluid phase.

For a phosphor-bronze mesh studied, the ratios of heat capacity and thermal conductivity between air and the mesh material are:

$$\beta = 4 \times 10^{-4} \text{ and } \lambda = 3 \times 10^{-4}$$

6.4.3 Time Constant

An important parameter describing the thermal performance of regenerators is the thermal time constant of the mesh wire, τ_c, defined as follows:

$$\frac{\alpha_f \tau_c}{d_w^2} = \frac{1}{\beta Nu_w} \qquad (6.19)$$

The time constant is the reciprocal of the parameter D discussed in Sect. 6.4.2. In non-dimensional form, with the tube radius as the reference dimension, we get

$$\frac{W\tau_c}{R} = \frac{d_w}{R}\frac{\text{Re Pr}}{\beta Nu} \qquad (6.20)$$

The non-dimensional characteristic frequency of the regenerator mesh can then be defined as

$$\omega_C = \frac{2\pi}{2\tau_c} = \frac{\pi}{\tau_c} \text{ rad/s} \qquad (6.21)$$

The quantity ω_C can be physically interpreted in terms of the cryocooler operation. It is the optimum frequency at which the device should be operated so that adequate time is available for heating the mesh in one part of the cycle and cooling it at the second stage. In this chapter, results are presented in terms of a frequency multiplier relative to the characteristic.

The dimensionless characteristic frequencies $R\omega_C/W$ are 0.177 and 0.021 for the mesh of [38] and 0.0157 and 0.00113 for the mesh of [23], respectively, at Re = 100 and 10,000. The mesh of Chen et al. [38] has a smaller wire diameter when compared to Sozen and Kuzay [23]. Naturally, it has a smaller time constant and larger characteristic frequency.

The thermal time constant can be compared to that of fluid flow (τ_F) as follows. The flow time constant can be estimated from dimensional analysis to fall in the range

$$\tau_F = \frac{d_w}{W} \text{ to } \frac{R}{W} \qquad (6.22)$$

Hence

$$\frac{\tau_F}{\tau_c} = \left[\frac{d_w}{R} \text{ to } 1\right]\frac{\beta Nu}{\text{Re Pr}}$$

Since $\beta \ll 1$ and Re $\gg 1$, the upper limit of the time constant ratio is much smaller than unity. Flow transients can, thus, be justifiably neglected in comparison with the thermal unsteadiness.

6.4.4 *Harmonic Analysis*

To solve the momentum equation, a harmonic analysis technique is used. First, the dimensionless cross-sectional average velocity $\overline{w(t)}$ is assumed to oscillate with unit amplitude at a non-dimensional frequency ω that matches the frequency of operation of the cryocooler. Thus (with real part implied)

$$\overline{w(t)} = \exp(i\omega t) \quad i = \sqrt{-1} \tag{6.23}$$

The velocity amplitude is unity and arises from using W as the velocity scale. The fully developed gas velocity distribution in the porous region is expanded in the following form:

$$w(r,t) = \sum_{j=1}^{N} \phi_j \exp(ij\omega t) \tag{6.24}$$

Equations 6.23 and 6.24 will apply when the device has reached a dynamic steady state. The extent to which flow transients will persist in the cryocooler is estimated in Sect. 6.4.3 and is shown to be small in real-life regenerators. The expectation from Eq. 6.24 is that the contribution from higher harmonics $(j > 1)$ will diminish for increasing values of the index j. In this respect, the series converges with an increasing order of the harmonic. This approach can be expected to be appropriate for flow in porous media even at higher Reynolds numbers for the following reason. We may expect higher harmonics arising from the nonlinearity of acceleration terms (in the form of $\mathbf{u} \cdot \nabla \mathbf{u}$) and that of form drag being proportional to $\mathbf{u}^2$. In contrast, the effect of introducing a stationary mesh in gas flow is to stabilize it in several aspects. Wall boundary layers are thin, and their transition to turbulence is delayed. Fluctuations in the gas phase are further damped in the core of the mesh. For this reason, the local velocity is expanded in multiples of the fundamental with the higher harmonics becoming progressively less important.

Functions ϕ_j can be viewed as velocity amplitudes that depend on the radial coordinate alone. The forcing frequency is generally known from the device operating characteristics, for example, that of the Stirling cryocooler. In the above equation, N represents the number of harmonics retained in the analysis. Results are presented in this chapter for up to $N = 4$; inclusion of higher modes was not found necessary for the studied conditions.

Noting that the pressure gradient in Eq. 6.5 is a constant with respect to the axial coordinate, one can expand pressure gradient also as:

$$-\frac{dp}{dz} = \sum_{j=1}^{N} P_j \exp(ij\omega t) \tag{6.25}$$

While φ_j is a complex function of r and P_j a complex constant, their real and imaginary parts must be determined as a part of the calculation. Substituting these expansions into Eqs. (6.4) and (6.5) and collecting terms of each order leads to the following equations:

$$\frac{\partial \phi_j}{\partial z} = 0, \quad j = 1 - 4 \tag{6.26}$$

$$\frac{i\omega}{\varepsilon} \phi_1 = P_1 - \frac{1}{ReDa} \phi_1 + \frac{1}{\varepsilon Re}(\phi_1'' + \frac{1}{r}\phi_1') \tag{6.27}$$

$$\frac{2i\omega}{\varepsilon} \phi_2 = P_2 - \frac{1}{ReDa} \phi_2 - \frac{F}{\sqrt{Da}}\phi_1^2 + \frac{1}{\varepsilon Re}(\phi_2'' + \frac{1}{r}\phi_2') \tag{6.28}$$

$$\frac{3i\omega}{\varepsilon} \phi_3 = P_3 - \frac{1}{ReDa} \phi_3 - \frac{2F}{\sqrt{Da}}\phi_1\phi_2 + \frac{1}{\varepsilon Re}(\phi_3'' + \frac{1}{r}\phi_3') \tag{6.29}$$

$$\frac{4i\omega}{\varepsilon} \phi_4 = P_4 - \frac{1}{ReDa} \phi_4 - \frac{F}{\sqrt{Da}}(\phi_2^2 + 2\phi_1\phi_3) \\ + \frac{1}{\varepsilon Re}\left(\phi_4'' + \frac{1}{r}\phi_4'\right) \tag{6.30}$$

Here, primes and double primes denote, respectively, the first and second derivatives of the corresponding modal velocity amplitude with respective to r. Specifically, we have

$$\phi' = \frac{\partial \phi}{\partial r} \quad \phi'' = \frac{\partial^2 \phi}{\partial r^2}$$

Equations 6.26–6.30 are coupled in the sense that pressure amplitudes in the momentum equation are initially unknown and must be determined from the condition of mass balance. Coupling between pressure and velocity can be resolved by employing the mass conservation equation

$$\frac{1}{\pi} \int_0^1 2\pi r \phi_j(r) dr = I, \quad j = 1 - 4 \tag{6.31}$$

Here, symbol I is 1 for $j = 1$ and is zero otherwise. Equation 6.31 enforces unit average velocity amplitude across the tube cross section, as required under fully developed flow conditions. Multiplying Eqs. (6.26) through (6.30) by $2\pi r$ and integrating as in Eq. 6.31, the following algebraic equations for the pressure gradients P_j can be obtained:

$$\frac{i\omega}{\varepsilon} = P_1 - \frac{1}{Re\,Da} + \frac{2}{\varepsilon\,Re}\,\phi_1'(1) \tag{6.32}$$

$$0 = P_2 - \frac{F}{\pi\sqrt{Da}} \int_0^1 2\pi r\phi_1^2 dr + \frac{2}{\varepsilon Re}\,\phi_2'(1) \tag{6.33}$$

$$0 = P_3 - \frac{F}{\pi\sqrt{Da}} \int_0^1 4\pi r\phi_1\phi_2 dr + \frac{2}{\varepsilon Re}\,\phi_3'(1) \tag{6.34}$$

$$0 = P_4 - \frac{F}{\pi\sqrt{Da}} \int_0^1 2\pi r(\phi_2^2 + 2\phi_1\phi_3)dr + \frac{2}{\varepsilon Re}\,\phi_4'(1) \tag{6.35}$$

Here, $\phi_j'(r = 1)$ can be interpreted as the dimensionless wall velocity gradient. Eliminating P_j of Eqs. (6.26)–(6.30) with Eqs. (6.32–6.35), differential equations for ϕ_j are obtained. The boundary conditions to solve the second-order differential Eqs. (6.26–6.30) of $\phi_j(r)$ are one of no-slip at the wall ($r = 1$) and symmetry at the axis ($r = 0$) and are stated as follows:

$$\text{at the wall } \phi_j\big|_{r=1} = 0 \tag{6.36}$$

$$\text{along the axis } \frac{\partial \phi_j}{\partial r}\bigg|_{r=0} = 0 \tag{6.37}$$

The first equation describes the no-slip condition at the wall and the second the axisymmetry of the flow. Separating real and imaginary parts of the complex-valued equations, both real and imaginary parts of φ_j can be determined. However, simultaneous solutions for each of them are still required in the solution process. This is because the gradient of ϕ_j at the wall appearing in Eqs. (6.32) through (6.35) is an unknown complex constant and should be determined simultaneously so as to fit the boundary conditions of the amplitude functions ϕ_j. Once the solution for velocity is obtained, the time-dependent pressure field can be calculated from Eqs. (6.32–6.35).

The real and imaginary parts of the momentum equations of orders 1–4 thus derived are listed below.

Order 1

REAL: $-\dfrac{\omega}{\varepsilon}\phi_{1i} = -\dfrac{1}{ReDa}(\phi_{1r} - 1) + \dfrac{1}{\varepsilon Re}\left(\phi_{1r}'' + \dfrac{1}{r}\phi_{1r}'\right) - \dfrac{2}{\varepsilon Re}\phi_{1r}'(1)$

IMAGINARY: $\dfrac{\omega}{\varepsilon}(\phi_{1r} - 1) = -\dfrac{1}{ReDa}\phi_{1r} + \dfrac{1}{\varepsilon Re}\left(\phi_{1i}'' + \dfrac{1}{r}\phi_{1i}'\right) - \dfrac{2}{\varepsilon Re}\phi_{1i}'(1)$

Order 2

$$\text{REAL:}\quad -2\frac{\omega}{\varepsilon}\phi_{2i} = -\frac{1}{Re\,Da}\phi_{2r} + \frac{1}{\varepsilon\,Re}(\phi''_{2r} + \frac{1}{r}\phi'_{2r}) - \frac{2}{\varepsilon\,Re}\phi'_{2r}(1)$$

$$-\frac{F}{\sqrt{Da}}(\phi^2_{1r} - \phi^2_{1i}) + \frac{2F}{\sqrt{Da}}\int_0^1 r(\phi^2_{1r} - \phi^2_{1i})dr$$

$$\text{IMAGINARY:}\ 2\frac{\omega}{\varepsilon}\phi_{2r} = -\frac{1}{Re\,Da}\phi_{2i} + \frac{1}{\varepsilon\,Re}(\phi''_{2i} + \frac{1}{r}\phi'_{2i}) - \frac{2}{\varepsilon\,Re}\phi'_{2i}(1)$$

$$-\frac{F}{\sqrt{Da}}(2\phi_{1r}\phi_{1i}) + \frac{2F}{\sqrt{Da}}\int_0^1 r(2\phi_{1r}\phi_{1i})dr$$

Order 3

$$\text{REAL:}\quad -3\frac{\omega}{\varepsilon}\phi_{3i} = -\frac{1}{Re\,Da}\phi_{3r} + \frac{1}{\varepsilon\,Re}(\phi''_{3r} + \frac{1}{r}\phi'_{3r}) - \frac{2}{\varepsilon\,Re}\phi'_{3r}(1)$$

$$-\frac{2F}{\sqrt{Da}}(\phi_{1r}\phi_{2r} - \phi_{1i}\phi_{2i}) + \frac{2F}{\sqrt{Da}}\int_0^1 2r(\phi_{1r}\phi_{2r} - \phi_{1i}\phi_{2i})dr$$

$$\text{IMAGINARY:}\ 3\frac{\omega}{\varepsilon}\phi_{3r} = -\frac{1}{Re\,Da}\phi_{3i} + \frac{1}{\varepsilon\,Re}(\phi''_{3i} + \frac{1}{r}\phi'_{3i}) - \frac{2}{\varepsilon\,Re}\phi'_{3i}(1)$$

$$-\frac{2F}{\sqrt{Da}}(\phi_{1i}\phi_{2r} + \phi_{1r}\phi_{2i}) + \frac{2F}{\sqrt{Da}}\int_0^1 2r(\phi_{1i}\phi_{2r} + \phi_{1r}\phi_{2i})dr$$

Order 4

$$\text{REAL:}\ -4\frac{\omega}{\varepsilon}\phi_{4i} = -\frac{1}{Re\,Da}\phi_{4r} + \frac{1}{\varepsilon\,Re}(\phi''_{4r} + \frac{1}{r}\phi'_{4r}) - \frac{2}{\varepsilon\,Re}\phi'_{4r}(1) - \frac{F}{\sqrt{Da}}((\phi^2_{2r} - \phi^2_{2i})$$

$$+ 2(\phi_{1r}\phi_{3r} - \phi_{1i}\phi_{3i})) + \frac{2F}{\sqrt{Da}}\int_0^1 r((\phi^2_{2r} - \phi^2_{2i}) + 2(\phi_{1r}\phi_{3r} - \phi_{1i}\phi_{3i}))dr$$

$$\text{IMAGINARY:}\ 4\frac{\omega}{\varepsilon}\phi_{4r} = -\frac{1}{Re\,Da}\phi_{4i} + \frac{1}{\varepsilon\,Re}(\phi''_{4i} + \frac{1}{r}\phi'_{4i}) - \frac{2}{\varepsilon\,Re}\phi'_{4i}(1) - \frac{2F}{\sqrt{Da}}((\phi_{2r}\phi_{2i})$$

$$+ (\phi_{1r}\phi_{3i} + \phi_{1i}\phi_{3r})) + \frac{2F}{\sqrt{Da}}\int_0^1 2r((\phi_{2r}\phi_{2i}) + (\phi_{1r}\phi_{3i} + \phi_{1i}\phi_{3r}))dr$$

The equations for ϕ_j are decoupled for each j, but require a simultaneous solution for the real and imaginary parts. The elimination of the pressure gradient introduces the wall velocity gradient as a new unknown, which must be determined as a part of the global solution. The numerical procedure adopted in the present study is to solve for the real and imaginary parts of velocity iteratively with corrections for the complex wall velocity gradients being determined from the mass balance equation by a Newton–Raphson scheme. Each two-point boundary value problem has been solved by central differencing of the derivatives followed by TDMA [41] for matrix inversion. A uniform grid of 401 points in the radial direction was found necessary to obtain a grid-independent solution. The algorithm for determining velocity and pressure amplitudes of each of the harmonics is as follows.

i. Assume the value of the complex wall velocity gradient of each order (1–4).
ii. Integrate the system of coupled second-order ODEs (real and imaginary parts of each order) from the wall to the axis as an initial value problem. Here, note that the velocity amplitude of a lower order appears as a source term in the ODE of the higher order.
iii. Examine if the gradient boundary condition at the axis is satisfied.
iv. Correct the wall velocity gradient using the Newton–Raphson scheme till the boundary condition at the axis is satisfied to within a prescribed tolerance.
v. The velocity field is now fully available in terms of the quantities $\phi_j(r)$ at the nodes of the radial mesh.
vi. Evaluate next the pressure amplitudes P_j from Eqs. (6.32–6.35).
vii. These calculations must be repeated for each frequency and Reynolds number, as well as for each mesh.

6.4.5 Numerical Solution of the Energy Equation

The solution of the energy Eqs. (6.6–6.7) is considered next. As discussed in Sect. 6.4.3, the assumption of a short-lived transient applicable for flow is no longer valid since the transient durations are determined by the thermal properties of the metal wire of the mesh and the pulsing fluid. For practical configurations of regenerators, the transient duration is substantially large. Hence, one is required to solve the full-time-dependent energy equations for the fluid and the solid. The passage of a pulse of hot fluid followed by one of cold fluid in the opposite direction constitutes one cycle of the regenerator operation. All thermal calculations are carried out in the present study for 10^4 cycles. At lower frequencies, the thermal field was found to have reached dynamic steady state at the end of this period. At higher frequencies, the time available for heating and cooling the mesh was small; consequently, the transients were uniformly slow and a strict dynamic steady state was not attained.

The flow field for temperature calculation is specified as:

$$w(r,t) = \text{Real}\left[\sum_{j=1}^{4} \phi_j \exp(ij\omega t)\right] \tag{6.38}$$

Energy equations for the fluid as well as the solid phases are solved by an implicit time-marching procedure with central differencing for the diffusion terms, the QUICK scheme [42] for the advection term with the solid phase temperature lagging the fluid phase temperature by one time step. The discretized equations are cast in the form of a successive substitution formula and solved by Gauss–Seidel iteration. All gradient boundary conditions are embedded into the finite difference form of the differential equation to ensure diagonal dominance. Temperature was found to be a constant along the radius in most calculations except the one which included heat losses to the ambient (Bi > 0). Hence, a relatively coarse mesh was used in the radial direction when the Biot number is zero and 10% of the tube radius for a Biot number of unity. In the flow direction, a typical grid size of 10% of the tube radius was found to be adequate for obtaining grid-independent results. The time step used is 1% of the time period of the base flow.

6.5 Results and Discussion

Results arising from the mathematical model of a regenerator in a Stirling cycle are presented in two parts: (a) flow characteristics of the regenerator mesh and (b) its thermal response. The flow behavior is presented in the form of velocity traces at selected locations along the tube radius and the friction factor. The distribution of the friction factor among the harmonics as well as the total equivalent value is presented. Here, the friction factor of the jth harmonic is equal to P_j and is a complex quantity. The total equivalent friction factor summed over all harmonics is defined as

$$f_{eq} = \sqrt{\sum_{j=1}^{4} \left| P_j^2 \right|} \qquad (6.39)$$

For a regenerator mesh of a cryocooler, its thermal response is of at most importance. The temperature profile across the regenerator should be linear at all time instants, diminishing from the source temperature to the sink. Specifically, one can ask how changes in the mesh length, porosity, frequency, and Reynolds number will help achieve the ideal temperature variation.

Quantities of interest while specifying the thermal response of the mesh are temperature profiles in the axial direction, variation of temperature with time at selected points in the mesh, extent of thermal equilibrium between the solid and the fluid phases, and the overall effectiveness of the regenerator. Effectiveness is defined in the present study as the ratio of the thermal energy of the fluid leaving the left plane (corresponding to a cold inflow condition on the right) to the total energy supplied to the mesh by the incoming hot fluid on the left plane during the positive half of the cycle (Sect. 6.5.3). The performance of the mesh has been studied as a function of frequency and Reynolds number. The frequency referred in discussions below is the frequency multiplier ω_M, which relates the pulsing frequency ω_M and the characteristic frequency ω_C through the expression, $\omega = \omega_M \times \omega_C$. For a discussion on the characteristic frequency, see Eq. 6.21 and the accompanying text.

6.5.1 Flow Behavior

The bulk of the results reported in the present study corresponds to the mesh tested by Chen et al. [38]. The mesh of Sozen and Kuzay [23] is also considered for comparison. The range of Reynolds numbers studied is 100–10,000, and the range of frequency multipliers (ω_M) is 1–1000, with respect to ω_C, the characteristic value. For the results presented, friction factor of the fourth harmonic is less than 1% of the leading harmonic, confirming that four harmonics are adequate to represent the flow field. Since the volumetric flow is prescribed as a sinusoidal function, local fluid velocities in the mesh are practically sinusoidal, except in the near-wall region. Hence, the role of the higher harmonics is visible in the friction factors.

The triggering of higher harmonics in the pressure field can be viewed as indicating the onset of instability of the base flow under oscillatory conditions. Instability is to be expected for the range of parameters considered in the study. For example, Da $\to \infty$ indicates the pure fluid limit at which Re = 10,000 represents turbulent flow. In contrast, a dense mesh (as in [38]) has a stabilizing influence and, at high Reynolds numbers, a well-behaved solution is possible. The physical mechanism responsible for stabilizing the flow within the regenerator is the formation of a very thin boundary-layer near the tube wall. As stated in Sect. 6.4.1, its thickness is proportional to $R\sqrt{Da}$ [15]. In general, hydrodynamic stability of internal flow through porous media is a topic of future research.

Figure 6.3a shows the velocity traces within a cycle for the mesh reported by Chen et al. [38]. The Reynolds number and frequency multiplier for this figure are 1000 and 100, respectively. Two radial positions, namely $r = 0$, the axis, and $r = 0.975$, very near the tube wall are shown. It can be seen that the velocity is near sinusoidal and there is negligible change in velocity in the radial direction all the way up to $r = 0.975$. These results have been equally observed in all calculations with the mesh of [38].

The friction factors as a function of Reynolds number and the frequency multiplier for the mesh under discussion are summarized in Table 6.1. In Tables 6.1, 6.2, and 6.3, symbol 'E + n' is the multiplier 10^n.

At Re = 100, the friction factor due to the leading harmonic dominates all others and almost completely accounts for the total friction factor. The leading harmonic is composed of the viscous pressure drop as the fluid moves past the solid wire and viscous friction between the bulk of the fluid and the bounding tube wall. The second and higher orders reflect perturbations arising from flow separation effects at the wire surface. These are clearly small at Re = 100 since the corresponding wire-diameter-based Reynolds number is only 0.1. At Re = 10,000, the real part of

Fig. 6.3 Time traces of
velocity at two locations in
the porous insert serving as a
regenerator. Re = 1000;
ω_M = 100. **a** Uniform
porosity dense mesh; **b** dense
mesh with porosity variation
near the wall; **c** coarse mesh

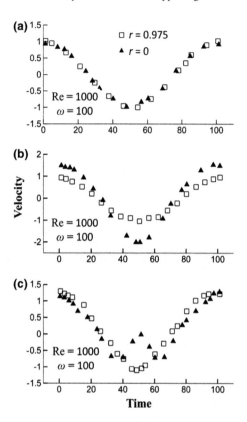

the friction factor of first order decreases by a factor of 100. The contributions of the
viscous and inertia effects are comparable, and the total friction factor has equal
contributions from the first and the second harmonics. The roles of the third and
fourth harmonics continue to be negligible.

The effect of increasing the pulsing frequency is to raise the magnitude of the
imaginary part of the friction factor. This results in a slight increase in the pressure
drop and can be interpreted just as the introduction of a phase difference for
pressure with respect to the flow field, which is close to sinusoidal. The sign of the
phase difference is positive. The phase difference is most dominant in the leading
harmonic and attains a maximum of 1.86° at Re = 100 and 21.9 degrees at
Re = 10,000, both at ω_M = 1000. The increase in phase lag with frequency at each
Reynolds number is practically linear over the frequency range considered. The
phase difference at Re = 10,000 is larger than at Re = 100, but continues to be
completely associated with the leading harmonic. Phase differences associated with
the third and fourth harmonics at Re = 10,000 are also large, but since their
respective amplitudes are small, they are not significant. These results suggest that
the origin of the phase lag between pressure drop and flow is analogous to that in
the forced oscillations of a massless spring-damper system, spatial variation due to
the tube wall, and inertial effects having only a secondary role.

Table 6.1 Real and imaginary parts of the friction factor P_j of four orders ($j = 1 \ldots 4$) as a function of Re and frequency (mesh of [38])

ω	Real (1)	Imag (1)	Real (2)	Imag (2)	Real (3)	Imag (3)	Real (4)	Imag (4)	f_{eq}
Re = 100									
1	7.70E + 03	0.25	2.10E + 02	0	−0.69E − 2	0	0	0	7.33E + 03
10	7.70E + 03	2.50E + 00	2.10E + 02	0	−0.69E − 2	0	0	0	7.73E + 03
100	7.70E + 03	2.50E + 01	2.10E + 02	0	−0.69E − 2	0	0	0	7.73E + 03
1000	7.70E + 03	2.50E + 02	2.10E + 02	−0.9E − 2	−0.7E − 2	0	0	0	7.73E + 03
Re = 10,000									
1	7.70E + 01	0.3E − 1	2.10E + 02	0	−0.7	0	2.60E + 00	−0.6E − 2	2.25E + 02
10	7.70E + 01	0.3	2.10E + 02	−0.11E − 2	−0.69	0.92E − 2	2.60E + 00	−0.59E − 1	2.25E + 02
100	7.70E + 01	3.10E + 00	2.10E + 02	−0.1E − 1	−0.68	0.92E − 1	2.60E + 00	−0.58	2.25E + 02
1000	7.70E + 01	3.10E + 01	2.10E + 02	−0.89E − 1	−0.35	0.44	1.10E + 00	−0.18E + 1	2.27E + 02

Table 6.2 Real and imaginary parts of the friction factor P_j of four orders ($j = 1\ldots4$) as a function of Reynolds number and frequency (mesh of [38]); effect of variable porosity near the wall

ω	Real (1)	Imag (1)	Real (2)	Imag (2)	Real (3)	Imag (3)	Real (4)	Imag (4)	f_{eq}
Re = 100									
1	7.90E + 03	0.27	2.50E + 02	−0.8E − 2	−0.75E + 1	0.33E − 2	1.80E + 00	−0.13E − 2	7.96E + 03
10	7.90E + 03	2.70E + 00	2.50E + 02	−0.8E − 1	−0.75E + 1	0.33E − 1	1.80E + 00	−0.13E − 1	7.96E + 03
100	7.90E + 03	2.70E + 01	2.50E + 02	−0.8	−0.75E + 1	0.33	1.80E + 00	−0.13	7.96E + 03
1000	7.90E + 03	2.70E + 02	2.50E + 02	−0.78E + 1	−0.65E + 1	3.00E + 00	1.30E + 00	−0.11E + 1	7.96E + 03
Re = 10,000									
1	7.90E + 01	0.3E − 1	2.50E + 02	−0.98E − 1	−0.75E + 3	4.10E + 00	1.80E + 04	−0.17E + 3	2.70E + 04

Table 6.3 Real and imaginary parts of the friction factor P_j of four orders ($j = 1 \ldots 4$) as a function of Re and frequency (mesh of [23])

ω	Real (1)	Imag (1)	Real (2)	Imag (2)	Real (3)	Imag (3)	Real (4)	Imag (4)	f_{eq}
Re = 100									
1	3.40E + 01	0.19E − 1	3.00E + 00	0	−0.6E − 2	0	0.55E − 3	0	3.40E + 01
10	3.40E + 01	0.19	3.00E + 00	−0.48E − 3	−0.6E − 2	0	0.55E − 3	0	3.40E + 01
100	3.40E + 01	1.90E + 00	3.00E + 00	−0.48E − 2	−0.6E − 2	0.85E − 3	0.55E − 3	−0.16E − 3	3.50E + 01
1000	3.40E + 01	7.50E + 00	3.00E + 00	−0.18E − 1	−0.46E − 2	0.27E − 2	0.21E − 3	−0.38E − 3	3.50E + 01
Re = 10,000									
1	0.346	0.13E − 2	3.00E + 00	−0.35E − 3	−0.61	0.62E − 2	5.50E + 00	−0.12	6.30E + 00

Results for the real and imaginary parts of the friction factor obtained in the present work can be analytically explained from the momentum equation to the leading harmonic written in the frequency domain (Eq. 6.27) as follows:

$$\frac{i\omega}{\varepsilon}\phi_1 = P_1 - \frac{1}{Re\,Da}\phi_1 + \frac{1}{\varepsilon Re}(\phi_1'' + \frac{1}{r}\phi_1')$$

Since the velocity variation in the mesh is practically uniform, the last bracketed term can be dropped, leading to the simplified form

$$\frac{i\omega}{\varepsilon}\phi_1 = P_1 - \frac{1}{Re\,Da}\phi_1$$

Since ϕ_1 across the cross section is mostly real and close to unity, this equation shows that the real part of the friction factor decreases inversely with Reynolds number and the imaginary part increases linearly with frequency. The phase difference of the pressure gradient with respect to the superimposed flow can be calculated as

$$\text{phase difference} \sim \tan^{-1}\left[\frac{\omega\,Re\,Da}{\varepsilon}\right] \tag{6.40}$$

When the phase difference is small, it increases linearly with frequency. For large values of the argument in Eq. 6.40, for example, the pure fluid limit $(Da \rightarrow \infty)$, phase difference approaches $\pi/2$ radians. Phase difference between oscillatory flow and pressure drop angles is generally larger in a pure fluid, when compared to dense meshes.

Mesh screens stacked together in a tube to form a regenerator will exhibit some heterogeneity in the spatial direction. Changes in porosity with the axial coordinate cannot be accommodated in a one-dimensional model for flow. The effect of variation of porosity near the tube wall can, however, be studied. In analogy with a packed bed of spherical particles, the porosity variation is taken in the form [17].

$$\varepsilon(y) = \varepsilon_0\left(1 + A_2 \exp\left(-A_1\frac{(y-1)}{d_w}\right)\right) \tag{6.41}$$

Here, the coordinate y is the distance from the tube wall. Constants A_1 and A_2 are determined in such a way that the porosity variation spans over 10 wire diameters and the maximum porosity appearing at the wall. An immediate consequence of variation in porosity is that permeability becomes a function of the radial coordinate, being large in regions of large porosity. This functionality is introduced in the numerical model as

$$Da = \frac{A_3 \varepsilon^3}{(1 - \varepsilon)^2} \tag{6.42}$$

The constant A_3 is determined by requiring

$$\int_0^1 2r Da(r) dr = Da_0 \tag{6.43}$$

The symbol Da_0 is the average Darcy number determined from pressure drop experiments during steady flow through the mesh. Values of A_1, A_2, and A_3 obtained in the present study for the mesh of [38] are 0.304, 0.2091, and 0.2281, respectively.

The velocity traces for the mesh including porosity variation near the wall are shown in Fig. 6.3b. Larger velocity amplitude near the tube wall compared to the axis arises from an increased porosity at that location. The corresponding friction factor data for this case is presented in Table 6.2. At Re = 100, the friction factor of the leading harmonic continues to dominate all other components. The imaginary parts of friction factors are equally small for all harmonics, showing that the pressure gradient and the flow fluctuate in phase with each other. The friction factor, including variable porosity, is marginally larger (=0.796E + 4) compared to constant porosity (=0.773E + 4). This increase is related to the additional pressure drop that takes place owing to a preferential flow path being established near the tube wall and is viscous in origin. A completely different behavior in friction factor is seen at Re = 10,000. The friction factors (mainly, the real part) are seen to increase substantially for all harmonics. The equivalent friction factor is no longer a reliable estimate at this Reynolds number since the series expansion in terms of harmonics has not converged at the fourth order. This is indicative of the possibility of instability of the base flow, mainly near the solid wall where the increase in porosity leads to the thickening of the boundary layer. The data for Re = 10,000 is presented here as representative of flows undergoing transition in a high porosity regenerator mesh. This data also shows that special attention should be paid to the meshes installed as regenerators in tubes. Chen et al. [38] have suggested the use of oversize screens to circumvent the problem of channeling of flow near the tube walls.

The second mesh that has been considered for analysis is the one described by Sozen and Kuzay [23]. The properties of this mesh are given in Sect. 6.4.2. This mesh has a higher porosity with respect to [38], a larger wire diameter, and correspondingly a lower characteristic frequency, ω_C. The velocity traces for this mesh are shown in Fig. 6.3c. The velocity trace at a point very near the wall shows the presence of higher harmonics. This trend was seen to persist only over a short distance near the tube wall, and the bulk of the flow was practically sinusoidal.

The friction factor data for the mesh of [23] is presented in Table 6.3. At Re = 100, it has properties similar to the first mesh (Table 6.1). Accordingly, the imaginary parts of the friction factor of all orders are equally small, the real part of the leading harmonic dominates others, and the imaginary part of the leading harmonic increases with frequency. The largest phase difference between the pressure gradient and the superimposed flow occurs at $\omega_M = 1000$ and is equal to $12°$. This is larger than the value seen for the mesh of [38]. Referring to the estimate of the phase tangent (Eq. 6.40) in dense meshes as $\omega \mathrm{Re}\, Da/\varepsilon$, one can associate the increase in phase angle with the increase in Darcy number. The proportionality between phase difference and frequency is seen here only for $\omega_M < 100$. Hence, at larger frequencies, viscous effects associated with the tube wall also play a role in determining the pressure gradient.

At $\mathrm{Re} = 10,000$, real parts of the friction factors are finite and do not diminish with increasing order of the harmonics. Thus, as for a mesh with porosity variation near the wall, flow in the mesh of [23] approaches instability. The operating conditions in Stirling cycles produce Reynolds numbers closer to 10,000. Hence, the conclusion that can be drawn from the present analysis is that meshes with a large porosity are not suitable as regenerators since they could be subjected to flow instability and a large pressure drop.

6.5.2 Thermal Performance

Thermal performance of the regenerator mesh is of primary importance in applications and is considered in detail in the present section. As stated earlier, all heat transfer calculations have been carried out for 10^4 cycles. While dynamic steady state was attained in some of the simulations, very slow transients were observed in a few others, particularly at higher forcing frequencies. The results obtained at the end of 10^4 cycles are presented first. Transient evolution of the temperature field is subsequently discussed.

Quantities of interest are the minimum and maximum temperatures attained by the solid and fluid phases, respectively, within a cycle, the regenerator effectiveness, and the variation of temperature with distance. Temperature variation is expected to be close to linear under ideal conditions, with minimal temperature deficit/excess at the source/sink side of the regenerator. These trends depend on the pulsing frequency, Reynolds number, porosity, heat losses to the ambient, and the length of the regenerator. Length does not appear in the flow calculations of Sect. 6.5.1 since velocity is taken as fully developed and friction factors are based on pressure drop on a unit length basis. It is however an important parameter for heat transfer within the regenerator.

6.5.3 Dynamic Steady State

From a heat exchanger viewpoint, it is desirable that the mesh effectiveness be as close to unity as possible, since it represents energy conservation, namely complete transfer of energy between the hot and the cold streams via the regenerator mesh. The requirements with respect to the temperatures are less clear. If only one plane of the mesh were to be employed, it would have to be heated to the source temperature in the first half of the cycle and cooled to the sink temperature in the second half. It is then clearly desirable that the temperature extremes attained by the mesh are as widely separated as possible. This indicates good thermodynamic performance. However, for meshes of finite length, each plane of the mesh is exposed to previously heated (or cooled) fluid. Hence, the heating and cooling of the fluid during each half of the cycle over the complete temperature range must be accomplished over the full length of the regenerator. One can require the increment in fluid temperature to be uniform as it goes from the cold to the hot reservoir or vice versa. This leads to the conclusion that the temperature profile through the ideal regenerator is linear. The temperature profile has a negative gradient in configurations where the heated fluid enters the regenerator on the left plane. In dimensionless terms, the midpoint of the regenerator should be at a temperature of 0.5 (fluid and solid), at all times. Jointly, the mesh effectiveness, suitably defined, must approach unity. Reasons for deviation from this value as a function of Reynolds number, frequency, and porosity are analyzed here.

Effectiveness of the regenerator mesh can be defined along the following lines. Let T_H and T_C be the source and sink temperatures across the regenerator. Let T_h be the temperature to which the fluid is heated up from T_C during one half of the cycle. In the next half of the cycle, it cools from T_H to T_c. Under dynamic steady-state conditions, the following equality holds:

$$T_h - T_C = T_H - T_c$$

The effectiveness of the regenerator can now be defined as the ratio

$$E = \frac{T_h - T_C}{T_H - T_C} \tag{6.44}$$

Effectiveness is a fraction (between 0 and unity) and may also be expressed as a percentage. Temperature T_h and T_c in Eq. 6.26 will depend on the radial coordinate. Since the flow distribution is uniform and wall losses to the ambient are negligible, temperature across the tube cross section can be taken to be spatially constant. These temperatures will fluctuate with time, and a cycle-averaged value can be used to evaluate effectiveness.

Table 6.4 summarizes thermal performance of the mesh reported in [38] under dynamically steady conditions. Reynolds number is 100, and the regenerator length is 5, relative to the tube radius. Temperatures in Table 6.4, as well as all other tables in this section, have been collected at the midpoint of the regenerator ($r = 0$; $z = L/2$) along the tube axis.

Observations based on Table 6.4 are as follows: The solid and fluid phase temperatures are close to each other over the range of frequencies (1–1000) studied. The solid wire temperatures are bounded by the fluid temperatures on the lower as

Table 6.4 Regenerator effectiveness and midpoint temperature for the mesh screen of [38] (Re = 100; L = 5)

ω	θ_f		θ_s		E
	max	min	max	min	
1	0.503	0.501	0.503	0.502	0.908
10	0.502	0.5	0.501	0.501	0.942
100	0.502	0.5	0.501	0.501	0.95
1000	0.214	0.213	0.213	0.213	0.932

well as the higher side. The difference between the maximum and the minimum temperatures is small, from a measurement point of view, and both are close to 0.5. Hence, the mesh is close to the optimum state, in the sense that the temperature profiles in the regenerator are practically linear. Deviation from the value of 0.5 is seen at $\omega_M = 1000$. As discussed below, the thermal transients last longer at higher pulsing frequencies and the lowered midpoint temperature is partly due to a steady state not having been reached, even after 10^4 cycles. The next observation from Table 6.4 is that effectiveness increases with frequency. The exception occurs at $\omega_M = 1000$ because a true dynamic steady state has not been reached. Since the midpoint temperature does not fall rapidly with frequency, the increase in effectiveness suggests that the mesh can be advantageously pulsed at frequencies greater than its characteristic value determined from the Reynolds number.

The effect of Reynolds number on the regenerator performance is considered next. Table 6.5 presents the solid and fluid phase temperatures and effectiveness at Re = 10,000, the regenerator length being $L = 10$. Thermal equilibrium between the solid and fluid phases continues to be seen here. There is initially a marginal increase in effectiveness with frequency, though a larger drop is seen at $\omega_M = 1000$. The midpoint temperatures deviate significantly from 0.5, being greater when ω_M is less than 10 and lower when it exceeds 100. At $\omega_M = 1000$, the midpoint temperature is barely raised above the cold fluid temperature.

The effect of increase in frequency on the regenerator performance (Tables 6.4 and 6.5) can be explained. The thermal capacity of the mesh is a weighted average of the fluid and the solid thermal capacities, the weights depending on the porosity. For values of porosity encountered in practice, the thermal capacity corresponds entirely to that of the solid phase (except at very low temperatures).

At a low frequency of pulsation (closer to the characteristic value given by Eq. 6.21), the temperature values of the solid as well as the fluid undergo a full cycle of heating and cooling, the average temperature at the center of the mesh approaching 0.5. As the frequency increases, a shorter time interval is available for the mesh temperature to rise to the previous level. This can be viewed as an increase in the equivalent thermal capacity of the mesh with frequency. Hence, the amplitude of the temperature fluctuations within the mesh will decrease with increasing frequency. Since the thermal capacity of the mesh corresponds almost entirely to the solid phase, the effect of increasing frequency is felt directly on the solid phase temperature. This trend is shown in Tables 6.4 and 6.5, particularly in Table 6.5 at Re = 10,000. Specifically, one sees the solid phase temperature approaching a constant with fluid temperatures oscillating about this value, though with a reduced

Table 6.5 Regenerator effectiveness and midpoint temperature for the mesh screen of [38] (Re = 10,000; $L = 10$)

ω	θ_f		θ_s		E
	max	min	max	min	
1	0.582	0.564	0.579	0.566	0.95
10	0.577	0.566	0.573	0.571	0.953
100	0.245	0.234	0.239	0.239	0.932
1000	0.6E − 4	0.4E − 4	0.5E − 4	0.5E − 4	0.847

amplitude. The damping of the solid phase temperature fluctuation indicates a larger instantaneous temperature difference between the fluid and solid phases, a larger interfacial heat flux, and so an enhanced effectiveness. This result is consistently seen when ω_M increases (from 1 to 100 at Re = 100 and from 1 to 10 at Re = 10,000), but at $\omega_M = 1000$, delayed transients persistently lower the effectiveness.

The dependence of the regenerator performance on Reynolds number can be understood in the following manner. As Re increases from 100 to 10,000, the interphase (fluid-to-solid) Nusselt number increases, the thermal front becomes sharper, and at a given plane of the mesh, a larger fraction of the heated fluid bypasses the solid wire. The characteristic frequency also shows a considerable increase with Reynolds number (Eq. 6.30). The increase in Nusselt number lowers the heat transfer resistance, and the solid wire temperatures match closely the fluid temperature fluctuations, particularly at $\omega_M = 1$. At higher frequencies, the increase in the equivalent thermal capacity lowers the temperature levels and increases the duration of the transients. The second and third factors, namely the formation of thermal fronts and the bypassing of the mesh wire by the fluid at higher Reynolds numbers, ensure that fluid temperatures in the core of the regenerator are above 0.5. The lowering of the fluid and wire temperatures at $\omega_M = 1000$ is substantially larger at Re = 10,000 mainly because the ω_C is larger and the associated transients are truly very long.

The effects of increasing the regenerator length on its thermal properties are presented in Table 6.6. The Reynolds number for this data is 10,000. Frequency multipliers considered are 1 and 100, since $\omega_M = 1000$ leads to long transients, lowered effectiveness, and an unacceptable reduction in the regenerator performance. Table 6.6 shows that effectiveness increases monotonically with length. It is clearly because of an increased ability of the regenerator to extract thermal energy from the flowing fluid and return it during the second half of the cycle. The increase, however, is only marginal beyond $L = 5$. There are other beneficial features to be seen at this length. For example, the effectiveness at $\omega_M = 100$ is seen to have reached the level of $\omega_M = 1$ at the end of 10^4 cycles. The regenerator performance, namely effectiveness and linearity in the temperature profile, is broadly insensitive to increase in frequency. Hence, $L = 5$ may be thought of as an optimum for a Reynolds number of 10,000 over a frequency range of 1–100 (all other properties being fixed).

Table 6.6 Regenerator effectiveness and midpoint temperatures for the mesh screen of [38]; Re = 10,000

ω_M	θ_f		θ_s		E
	max	min	max	min	
L = 1					
1	0.573	0.451	0.561	0.464	0.768
100	0.552	0.470	0.512	0.511	0.794
L = 5					
1	0.553	0.525	0.55	0.528	0.929
100	0.486	0.467	0.477	0.477	0.927
L = 10					
1	0.582	0.564	0.579	0.566	0.95
100	0.245	0.234	0.239	0.239	0.932
L = 20					
1	0.648	0.641	0.648	0.642	0.963
100	0.7E − 2	0.6E − 2	0.7E − 2	0.7E − 2	0.928

The extreme fluid and solid temperatures show an interesting variation with length. For short lengths, the mesh wire does not absorb all the energy from the hot fluid (and conversely, does not fully heat the cold fluid). Correspondingly, the maximum midpoint temperature is greater than and the minimum is lower than 0.5. Since the Reynolds number is high, the solid phase temperature follows the oscillations of the fluid phase, when $\omega_M = 1$. At higher frequencies, the damping of temperature fluctuations in the solid phase limits the amplitude of the fluid phase temperature as well and temperature levels start to fall. At $L = 5$, the two factors are seen to be in balance and midpoint temperatures are in the vicinity of 0.5, justifying its choice as the optimum length. At higher lengths, the heated fluid is completely cooled during the first half of the cycle over a fraction of the regenerator length. The remaining part of the regenerator is unused, and the fluid here is cold. The length of the underutilized portion of the regenerator increases with frequency. This phenomenon plays an important role in determining the midpoint temperature. Hence, for $L = 10$ and 20 and $\omega_M = 1$, the midpoint temperatures are much larger than 0.5. When ω_M increases to 100, the two ends of the regenerator essentially determine its performance and the temperatures in the central core remain closer to their initial values, i.e., zero in the present study.

The dynamic steady-state temperature variation as a function of the axial coordinate at a certain instant in each half of the cycle at Re = 10,000 and $L = 5$ is shown in Fig. 6.4a. In the first half of the cycle, flow is from left to right, while in the second half, it is from right to left. The temperature profiles for each half are nearly stationary, practically overlap, and have a negative gradient for all time. However, in Fig. 6.3a–b, the sign of the gradient of the temperature profile for the second part of the cycle has been reversed, to preserve visual clarity. Hence, the arrows for both parts of the cycle point in the same direction. When the flow is from left to right, the incoming fluid is cooled by the regenerator and its temperature decreases in the flow direction. When the flow direction is reversed, the incoming cold fluid is heated by the regenerator (from $z/L = 0$ to 1) and the fluid recovers the source temperature of unity. During both halves of the cycle, the temperature distribution in the regenerator is close to linear, except for minor fluctuations, to an extent revealed in Tables 6.4–6.5. The effect of increasing frequency is to lower temperatures marginally, and this is also visible in Fig. 6.4a. Figure 6.4b shows the corresponding plot for $L = 10$. Deviation from linearity is more pronounced here showing departure of the regenerator performance from the optimum. The midpoint temperature is greater than 0.5 at ω_M equal to 1 and less than 0.5 at 100, for reasons discussed in the earlier paragraph.

6.5.4 Heat Losses

It is important to assess the sensitivity of the regenerator performance to external parameters. One such factor is heat loss to the ambient. For the present discussion, the ambient is taken to be at the cold fluid temperature. The tube holding the regenerator mesh could be made of stainless steel, but, for simplicity in specifying parameters, it is taken as the mesh material, phosphor-bronze. The tube diameter is 19 mm, and the corresponding heat transfer coefficient falls in the range of 1–10 W/m²-K. A representative value of $Bi = 3 \times 10^{-3}$ has been employed for the present calculation.

Fig. 6.4 Temperature profile along the axis of the tube carrying a porous insert; **a** Re = 10,000, $L = 5$; **b** Re = 10,000; $L = 10$. Mesh considered is of Chen et al. [38]

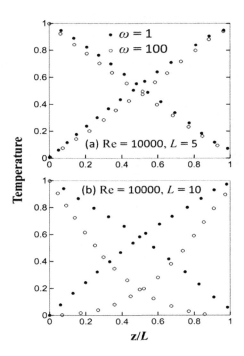

The regenerator properties including heat losses are summarized in Table 6.7. A visible drop in effectiveness as well as solid and fluid temperatures can be seen in this data. There is a marginal increase in effectiveness with increase in frequency, but the overall deterioration in the regenerator performance is dominant. One of the factors leading to a substantial heat loss is the dispersion in the transverse direction, given by the effective thermal conductivity in Eq. 6.18. Hence, one can conclude that the regenerator performance is quite sensitive to heat losses to the environment.

6.5.5 Coarse Mesh

Results given above were obtained using properties of the mesh reported by Chen et al. [38]. For comparison, the second mesh reported by [23] has also been analyzed. The main difference between the two meshes is in terms of porosity and wire diameter, both of which are larger for the mesh of [23]. Hence, the interfacial area A_{IF} is smaller, compared to [38]. The thermal properties of the regenerator made with this mesh are presented in Table 6.8. The steady-state temperature profiles plotted as a function of the axial coordinate are shown in Fig. 6.5.

The effectiveness for the coarser mesh is substantially smaller than for the denser one considered earlier, a result of an increase in porosity, and hence a drop in the equivalent thermal capacity. The Nusselt numbers are larger in the present case, but the interfacial area is much smaller, thus accounting for a large temperature fluctuation in the solid and fluid phase temperatures and a clearer demonstration of thermal nonequilibrium. Other results such as an increase in effectiveness with frequency, solid phase temperatures bounded by the corresponding fluid phase values and a reduction in the amplitude of the temperature fluctuation with increasing frequency, continue to be seen for the present mesh.

Table 6.7 Regenerator effectiveness and midpoint temperature for mesh screen of [38] in the presence of heat losses to the ambient (Re = 10,000, $L = 10$, $Bi = 3\times10^{-3}$)

ω	θ_f		θ_s		E
	max	min	max	min	
1	0.518	0.49	0.515	0.494	0.887
10	0.513	0.495	0.505	0.502	0.892
100	0.462	0.444	0.453	0.452	0.887
1000	0.2E − 1	0.2E − 1	0.2E − 1	0.2E − 1	0.805

Table 6.8 Regenerator effectiveness and midpoint temperature for mesh screen of [23] (Re = 10,000, L = 5)

Ω	θ_f		θ_s		E
	max	min	max	min	
1	0.84	0.188	0.766	0.271	0.616
100	0.746	0.268	0.514	0.508	0.669

Fig. 6.5 Steady-state temperature profile along the axis of the tube containing the porous insert; Re = 10,000, L = 5; mesh of Sozen–Kuzay [23]

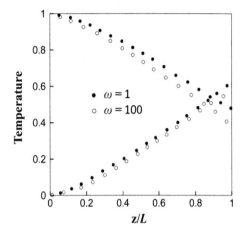

6.5.6 *Transient Response*

The thermal properties of the mesh under transient conditions are considered next. Figure 6.6a–c shows the variation of midpoint and average temperatures as a function of the cycle number. The temperatures shown are the average of the maximum and minimum values prevailing in the cycle. Figure 6.6a–c shows only the variation of the instantaneous fluid temperature. Solid phase temperatures show identical trends, except for a small change in their magnitudes. It is clear that the transient duration increases with an increase in frequency. In Fig. 6.6a (Re = 100), a clear dynamic steady state is reached even at $\omega_M = 100$. Further, the average and the midpoint temperatures become nearly equal at dynamic steady state. This can be interpreted as the appearance of a linear temperature profile through the regenerator. In fact, one can see the gradual increase in the average and midpoint temperatures with the cycle number till they eventually become equal. This indicates the movement of a thermal front (hot from the left side and cold from the right) till it fills the regenerator, marking the arrival of a dynamic steady state. In Fig. 6.6b, the Reynolds number is higher (=10,000). A dynamic steady state is not seen for $\omega_M = 100$. A dynamic steady state is reached at $\omega_M = 1$, the midpoint temperature

Fig. 6.6 Variation of
midpoint and average fluid
temperatures with time;
a Re = 100, L = 10;
b Re = 10,000, L = 10;
c L = 5; mesh of Chen et al.
[38]

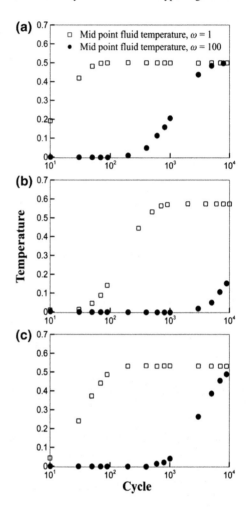

being greater than the average temperature. This is consistent with Fig. 6.4b where
a front-like temperature profile is formed. Figure 6.6c shows that the drawback can
be rectified with a regenerator of smaller length (L = 5) at the same Reynolds
number. In this case, a steady state is seen even at ω_M = 100 and the midpoint and
average temperatures approach each other after sufficient time has elapsed.

The variation of the effectiveness with cycle number at a Reynolds number of
10,000 is shown in Fig. 6.7a–d for four different regenerator lengths. Results for
ω_M = 10 have not been shown since they lie between those for ω_M = 1 and 100.
When ω_M = 1000, the regenerator performance evolves very slowly with time and

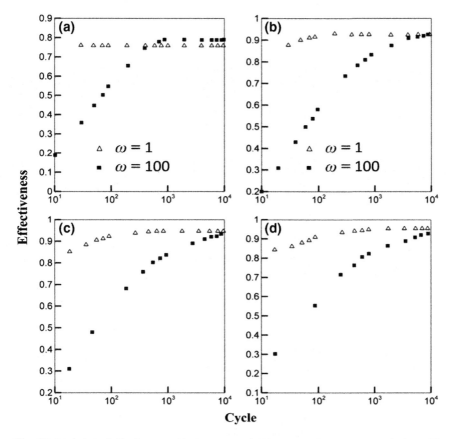

Fig. 6.7 Variation of effectiveness with time; Re = 10,000, **a** $L = 1$; **b** $L = 5$; **c** $L = 10$; **d** $L = 20$; mesh of Chen et al. [38]

effectiveness levels are small. The implication is that regenerators operating under these conditions will recover from unsteadiness intrinsic to engineering devices, quite slowly. Thus, the effectiveness levels recorded during the operation of the device will be far below the dynamic steady-state value. It is unlikely that the regenerator will be operated under these conditions.

For all lengths considered, the regenerator performance (Fig. 6.7) is adequate (in terms of reaching dynamic steady state quickly and the magnitude of effectiveness), when the frequency multiplier is unity. Differences arise only when the frequency takes on higher values. The shortest transients are shown in Fig. 6.7a, for the smallest regenerator ($L = 1$), for both $\omega_M = 1$ and 100. The increase in the long-term effectiveness with frequency is most clearly visible here. Figure 6.7b shows the transient effectiveness data for $L = 5$. The transients are longer-lived, but after 10^4 cycles, the effectiveness values are nearly equal for both frequencies considered.

There is only a marginal increase in steady-state effectiveness for lengths greater than 5. This supports the earlier conclusion that $L = 5$ is an optimum in the sense that the regenerator effectiveness continues to remain high over a range of frequencies. When $L = 10$ and 20, the transients are long-lived and the effectiveness levels at $\omega_M = 100$ are lower than at $\omega_M = 1$. Hence, these lengths represent suboptimum conditions for regenerator design.

6.6   Conclusions

A stack of meshes assembled to form a regenerator of a Stirling cycle has been analyzed for determining flow distribution and heat transfer. A non-Darcy, thermal nonequilibrium model is driven by pulsatile flow with hot and cold fluid alternately going past the mesh in opposite directions. The characteristic frequency of pulsation is determined with reference to the time constant of the gas-wire system. The flow is shown to reach dynamic steady state rapidly, when compared to the thermal field. With this result, the flow field is computed by a harmonic analysis technique. The energy equations for the fluid and the solid phases are numerically integrated in time. Calculations continue for a total of 10^4 cycles. Dense meshes of [38] are evaluated against the coarse meshes of [23]. Results have been presented for two Reynolds numbers, 100 and 10,000, the latter being closer to a practically attainable value in applications. Frequency multipliers from 1–1000 with respect to the characteristic value have been considered.

The following conclusions are drawn in the present work.

1. The velocity profiles that are established in the mesh of [38] are nearly flat, with small changes taking place near the tube wall. The velocity distribution is seen locally to be stable and in phase with the superimposed sinusoidal flow. The pressure gradient develops a phase difference with respect to the flow, but this is generally small. Inertia effects are significant at the higher Reynolds number studied. The mesh performance deteriorates when porosity variation near the tube wall is taken into account. The mesh of [23] also showed flow instabilities at the higher Reynolds number.
2. At a low Reynolds number, the steady-state effectiveness increases with frequency. At a higher Reynolds number, dynamic steady state was not reached even after 10^4 cycles.

3. For the mesh of [38], the solid and fluid phase temperatures were close to each other over a wide range of operating conditions. At a Reynolds number of 10^4, a regenerator length of $L = 5$ was found to be an optimum in the following respect. The regenerator effectiveness is high, the axial temperature profile in the flow direction is linear, and a stable performance is achieved over a frequency multiplier range of 1–100. The midpoint temperature oscillations were small and centered around 0.5.
4. For the mesh of [23], a higher porosity and a higher wire diameter resulted in a lower effectiveness and large temperature amplitude within a cycle. Thermal equilibrium was seen to be clearly violated. The lowered effectiveness made it unsuitable as a regenerator mesh.
5. A frequency multiplier of 1000 was found to be entirely unsuitable for the regenerator. The reasons are long thermal transients and reduced values of the mesh effectiveness.

References

1. G. Walker, *Cryocoolers Part I and II International Cryogenics Monographs Series* (Plenum Press, New York, 1983)
2. B. Cullen, J. McGovern, Development of a theoretical decoupled Stirling cycle engine. Simul. Model. Pract. Theory **19**, 1227–1234 (2011)
3. F. Formosa, G. Despesse, Analytical model for Stirling cycle machine design. Energy Convers. Manag. **51**, 1855–1863 (2010)
4. P.C.T. de Boer, Optimal regenerator performance in Stirling engines. Int. J. Energy Res. **33**, 813–832 (2009)
5. A.J. Organ, *The Air Engine, CRC Press, Boca Raton (2007); also see Stirling and Pulse-tube Cryocoolers* (Professional Engineering Publication Ltd, UK, 2005)
6. J.G. Brisson, G.W. Swift, Measurements and modeling of recuperators for superfluid Stirling refrigerator. Cryogenics **34**, 971–982 (1994)
7. L. Bauwens, Stirling cryocooler model with stratified cylinders and quasisteady heat exchangers. J Thermophys. Heat Trans. **9**, 129–135 (1995)
8. K. Muralidhar, K. Suzuki, Analysis of flow and heat transfer in a regenerator mesh using a non-Darcy thermally non-equilibrium model. Int. J. Heat Mass Trans. **44**, 2493–2504 (2001)
9. J. Tian, T.J. Lu, H.P. Hodson, D.T. Queheillalt, H.N.G. Wadley, Cross flow heat exchange of textile cellular metal core sandwich panels. Int. J. Heat Mass Trans. **50**, 2521–2536 (2007)
10. R. Dyga, M. Płaczek, Efficiency of heat transfer in heat exchangers with wire mesh packing. Int. J. Heat Mass Trans. **53**, 5499–5508 (2010)
11. J. Xu, J. Tian, T.J. Lu, H.P., Hodson, On the thermal performance of wire-screen meshes as heat exchanger material. Int. J. Heat Mass Trans. **50**, 1141–1115 (2007)
12. Chen Li, G.P. Peterson, The effective thermal conductivity of wire screen. Int. J. Heat Mass Trans. **49**, 4095–4105 (2006)
13. R.E. Hayes, A. Afacan, B. Boulanger, An equation of motion for an incompressible Newtonian fluid in a packed bed. Trans. Porous Media **18**, 185–198 (1995)
14. T. Masuoka, Y. Takatsu, Turbulence model for flow through porous media. Int. J. Heat Mass Trans. **39**, 2803–2809 (1996)

15. K. Vafai, S.J. Kim, Forced convection in a channel filled with a porous medium: An exact solution. Trans. ASME J. Heat Trans. **111**, 1103–1106 (1989)
16. G. Lauriat, V. Prasad, Non-Darcian effects on natural convection in a vertical porous enclosure. Int. J. Heat Mass Trans. **32**, 2135–2148 (1989)
17. K. Vafai, S.J. Kim, On the limitations of the Brinkman-Forchheimer-extended Darcy equation. IJHFF **16**, 11–15 (1995)
18. D.A. Nield, S.J. Kim, K. Vafai, Closure statements on the Brinkman-Forchheimer-extended Darcy model. IJHFF **17**, 34–35 (1996)
19. J.L. Lage, On the theoretical prediction of transient heat transfer within a rectangular fluid-saturated porous medium enclosure. ASME J. Heat Trans. **115**, 1069–1071 (1993)
20. B.V. Antohe, J.L. Lage, Amplitude effect on convection induced by time-periodic horizontal heating. Int. J. Heat Mass Trans. **39**, 1121–1133 (1996)
21. J. Bear, Y. Bachmat, *Introduction to Modeling of Transport Phenomena in Porous Media* (Kluwer Academic Publishers, London, 1990)
22. M. Kaviany, *Principles of Heat Transfer in Porous Media, Mechanical Engineering Series* (Springer, New York, 1991)
23. M. Sozen, T.M. Kuzay, Enhanced heat transfer in round tubes with porous inserts. Int. J. Heat Fluid Flow **17**, 124–129 (1996)
24. S.Y. Kim, B.H. Kang, J.M. Hyun, Heat transfer from pulsating flow in a channel filled with porous media. Int. J. Heat Mass Trans. **37**, 2025–2033 (1994)
25. A.V. Kuznetsov, An investigation of a wave of temperature difference between solid and fluid phases in a porous packed bed. Int. J. Heat Mass Trans. **37**, 3030–3033 (1994)
26. A.V. Kuznetsov, Comparison of the waves of temperature difference between the solid and fluid phases in a porous slab and a semi-infinite porous body. Int Comm. Heat Mass Trans. **22**, 499–506 (1995)
27. A.V. Kuznetsov, A perturbation solution for a nonthermal equilibrium fluid flow through three dimensional sensible storage packed bed. Trans. ASME J. Heat Trans. **118**, 508–510 (1996)
28. A.V. Kuznetsov, K. Vafai, Analytical comparison and criteria for heat and mass transfer models in metal hydride packed beds. Int. J. Heat Mass Trans. **38**, 2873–2884 (1995)
29. M. Sozen, K. Vafai, Analysis of the non-thermal equilibrium condensing flow of a gas through a packed bed. Int. J. Heat Mass Trans. **33**, 1247–1261 (1990)
30. K. Vafai, M. Sozen, Analysis of energy and momentum transport for fluid flow through a porous bed. Trans. ASME J. Heat Trans. **112**, 690–699 (1990)
31. A. Amiri, K. Vafai, Analysis of dispersion effects and non-thermal equilibrium, non-Darcian, variable porosity incompressible flow through porous media. Int. J. Heat Mass Trans. **37**, 939–954 (1994)
32. E.A. Foumeny, P.J. Heggs, Thermal performance of cyclic regenerators under varying inlet temperature conditions. Exp. Heat Trans. **8**, 1–15 (1995)
33. S. Isshiki, A. Sakano, I. Ushiyama, N. Isshiki, Studies on flow resistance and heat transfer of regenerator wire meshes of Stirling engines in oscillatory flow. Trans. JSME B **62–604**, 4254–4261 (1996)
34. S.Y. Kim, B.H. Kang, J.M. Hyun, Heat transfer from pulsating flow in a channel filled with porous media. Int. J. Heat Mass Trans. **37**, 2025–2033 (1994)
35. F. Kuwahara, A. Nakayama, H. Koyama, A numerical study of thermal dispersion in porous media. ASME J. Heat Trans. **118**, 756–761 (1996)
36. F. Kuwahara, A. Nakayama, H. Koyama, A study of thermal dispersion in fluid flow and heat transfer in porous media. Trans JSME B **62–600**, 3118–2124 (1996)
37. F. Kuwahara, A. Nakayama, H. Koyama, Direct numerical simulation of fluid flow in an anisotropic porous medium. Trans. JSME B **62–603**, 3920–3925 (1996)

38. P.-H. Chen, Z.-C. Chang, B.-J. Huang, Effect of oversize in wire screen matrix to the matrix holding tube on regenerator thermal performance. Cryogenics **36**, 365–372 (1996)
39. L.M. Jiji, *Heat Convection*, 2nd edn. (Springer-Verlag, Berlin, 2010)
40. A. Bejan, *Convection Heat Transfer*, 4th edn. (Wiley, New Jersey, 2013)
41. L.H. Thomas, Elliptic problems in linear difference equations over a network. Watson Sci. Comput. Lab. Rep. Columbia University, New York (1949)
42. B.P. Leonard, A stable and accurate convective modeling procedure based on quadratic upstream interpolation. Comput. Meth. Appl. Mech. Engg. **19**, 59–98 (1979)

Chapter 7
Geological Systems, Methane Recovery, and CO$_2$ Sequestration

Malay K. Das and Abhishek Khetan

7.1 Introduction

Geological porous media encompass a wide variety of naturally occurring structures including sand, clay, and rocks. Such structures play crucial roles in holding and transporting natural resources such as petroleum, natural gas, and groundwater. Furthermore, once depleted of natural resources, such geological reservoirs may be used for storing a variety of industrial by-products including CO$_2$. Understanding transport phenomena in geologic porous media is, therefore, essential from an energy and environmental viewpoint [1].

Despite significant efforts spanning several decades, the present knowledge on transport in geologic media has remained inadequate. The primary reason of such partial understanding is the complexities associated with the modeling of transport in geologic media. The geologic porous media constitute a multiscale problem where the pore scale ranges from few nm to mm, while the characteristic length-scale of the overall system is almost always in km [2]. The media often exhibit intermediate scales, generated by the presence of randomly oriented fractures. The overall system length-scale being quite large, the media are often presented with structural heterogeneity. In addition to the scaling issues, most practical applications in geologic systems involve multiphase–multispecies transport. The interfacial phenomena between the phases as well as the interaction of various phases with the solid matrix engender additional nonlinearities in the system [8]. Further, transport in geologic media often involve slow and sustained reactions or phase change leading to complex coupling of the balance equations. Finally, experimental results from a geologic media, usually derived from expensive seismic survey, provide limited information [3].

While the above characteristics prevail in almost all the geologic porous media, such media additionally display significant variability in structural and transport parameters. For instance, porosity and permeability in a tight gas reservoir are several orders of magnitude lower from a conventional natural gas reservoir [4].

© Springer International Publishing AG 2018
M.K. Das et al., *Modeling Transport Phenomena in Porous Media with Applications*, Mechanical Engineering Series,
https://doi.org/ 10.1007/978-3-319-69866-3_7

Both the reservoirs again behave differently than an artificially fractured shale gas well. In the present chapter, we focus on one particular unconventional energy resource, namely, marine hydrate reservoir. Such reservoirs are usually located at a high-pressure, low-temperature environment, deep beneath the sea. These reservoirs spread all over the globe near the coastal belt and usually suffer from low porosity and permeability. Marine hydrate reservoirs are identified as the largest storehouse of global carbon that is stored here as methane. While the high-pressure, low-temperature environment facilitates large storage of methane, the same environment forces the methane to bond with water to form clathrate hydrate.

Clathrate, adopted from Latin, means *endowed with crystal*. Clathrate hydrate, also known as gas hydrate, refers to an ice-like crystalline structure where guest gas molecules are encaged in a H_2O lattice [5]. The hydrate lattice, created by the hydrogen-bonded water molecules, shows three distinct structural variations: structure I, structure II, and structure H. The structures vary from each other according to the size of the cavity and the geometry of the lattice faces. The size of the cavity, formed inside the lattice, significantly influence the choice of the encaged gas. In naturally occurring hydrates, typical guest gas molecules include methane, ethane, propane, carbon dioxide, sulfur dioxide and hydrogen sulfide [6]. The unit lattice may also contain multiple cavity sizes leading to the formation of mixed gas hydrates. Detailed description of the hydrate structures is available elsewhere in the literature. Among the various naturally occurring hydrates, most important are the CH_4- and CO_2-hydrate [7–9]. The first, as indicated earlier, is identified as a potential energy resource, while the latter shows way for long-term storage of greenhouse gases. The CH_4-hydrate research delves into several important areas of energy and environment [10]. Some of them are briefly described below.

7.1.1 Gas Hydrate as an Energy Resource

Natural gas hydrate, primarily the hydrate form of CH_4, usually forms in low-temperature and high-pressure conditions in the presence of sufficient water and organic carbon-based compounds. CH_4 hydrates are routinely found under the seafloor at depths exceeding 300 m and in the arctic permafrost regions [11]. Current estimate shows that about 99% of natural gas hydrates occur in oceanic sediments while the rest stays in the permafrost hydrate reserves. Figure 7.1 shows the widespread distribution of hydrate deposits along the coastal lines, as obtained from a seismic survey [11]. While seismic measurements cannot accurately quantify the global CH_4 reserve in hydrate form, conservative estimate suggests that the reserve to be in the range of 10^{14}–10^{18} m^3 at STP [12]. Such an estimate, as shown in Fig. 7.2, indicates that the energy storage in CH_4 hydrates greatly surpasses the same in conventional fossil fuels [13]. Successful recovery of CH_4 from its hydrate may, therefore, greatly alter the global energy landscape in the coming years.

Fig. 7.1 Hydrate deposits along the coastal lines, as obtained from seismic survey

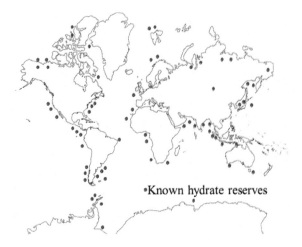

•Known hydrate reserves

Fig. 7.2 Global carbon balance in units of 10^{15} kg

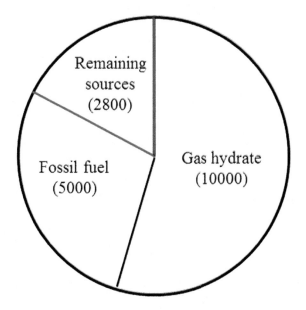

7.1.2   Environmental Concerns and CO_2 Sequestration

Sequestration of CO_2 in geologic formations is a technologically and financially viable option for controlling CO_2 concentration in the ambient atmosphere. The present discussion involves sequestration of CO_2 in marine hydrate reservoirs, which are the repositories of naturally formed CH_4-hydrates. While CH_4-hydrate consti-tutes the largest resource for CH_4 fuel, recovery of CH_4 from the marine hydrate reservoir seems to be quite challenging [14–16]. Injection of CO_2 in the CH_4-hydrate reservoir enhances CH_4 recovery via the formation of CO_2-hydrate. Sequestration of

CO_2 in its hydrate form ensures safe and long-term storage of CO_2 in large quantities. Finally, the sequestration of CO_2-hydrate in marine hydrate reservoir helps maintain the structural integrity of the fragile marine ecosystem [17–20].

7.1.3 Nature of Marine Hydrate Reservoirs and CH_4 Recovery

Figure 7.3 shows a typical hydrate reservoir in marine environment [21]. The hydrate reservoir is typically bounded by the impermeable seafloor. While many hydrate reservoirs are bounded at the bottom by impermeable rock structure, free methane and water may also be present below the hydrate deposit [22]. Based on how the hydrate deposit is bounded below, the accumulations can be divided into the following three main classes:

a. Class 1 accumulations comprise two layers: the hydrate layer (generally exhibiting very low permeability due to the presence of large hydrate saturation in pore space) and an underlying two-phase zone with free gas and water. In these accumulations, the bottom of hydrate stability zone (BHSZ) typically coincides with the bottom of the hydrate layer as with increasing depth the temperature also increases due to the geothermal temperature gradient [23–25]. This is the most desirable class for exploitation in terms of production, because of the hydrate thermodynamic proximity to hydration equilibrium, therefore necessitating only small changes in pressure and temperature to induce dissociation.
b. Class 2 accumulations comprise a hydrate-bearing layer overlying a mobile water zone (like an aquifer) with no free gas [26].
c. Class 3 accumulations are made up of a single hydrate layer bounded generally by impermeable under- and over-burdens. Such accumulations are characterized by the absence of an underlying zone of mobile fluids [27].

Fig. 7.3 Schematic of marine hydrate deposit; condition below the hydrate deposit defines the nature of reservoir

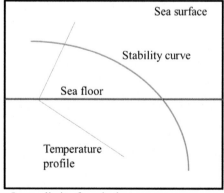

Lower limit of gas hydrate

Gas production from hydrate reservoirs can be classified into two broad schemes. The first is based on the idea of gas release by dissociation of hydrates. These schemes aim at altering the reservoir conditions so that the hydrate is moved outside its stability region, shown in Fig. 7.3. Three different production methods in this class are considered: (1) depressurization, (2) thermal stimulation, and (3) inhibitor injection [1–4, 28–31].

In depressurization, the gas phase pressure around the hydrate lattice is reduced across the hydrate stability zone leading to its dissociation. This can be achieved by removing gas and water from the reservoir through low-pressure extraction wells. Hydrates in the reservoir may not immediately start dissociation because the reservoir pressure may still be quite higher than the hydrate dissociation pressure at a given reservoir temperature. Since hydrate dissociation is an endothermic reaction, a decrease in reservoir temperature is likely to follow.

In the thermal stimulation technique, a hot fluid such as hot water is injected into the reservoir. The process relies on raising the reservoir temperature to destabilise the hydrate. This technique has been deemed inefficient because a significant portion of the energy introduced into the system is lost in the injection path and surroundings and only a fraction of injected energy is utilized toward the dissociation of gas hydrates.

The inhibitor injection method uses chemicals such as methanol to destabilize the hydrate eventually dissociating the CH_4. This technique seems to be expensive and therefore uneconomical for producing natural gas as large amounts of the inhibitor are required. Moreover, low porosity and relative permeability in hydrate formations may render it impossible to inject inhibitors.

Both depressurization and thermal stimulation have been applied at a field scale to assist production from hydrate deposits at Mallik Production Site (Mackenzie Delta, Canada). Early production tests in 2002 used thermal stimulation and showed a gas recovery of 470 Sm^3 (standard m^3). In 2007, a new production test using depressurization method was conducted and a total of gas output of 830 Sm^3 was obtained, but the tests were halted due to certain technical difficulties and sand accumulation. In 2008, continuous production for almost six days was achieved after some modifications during which gas flow rates ranging from 2000 to 4000 Sm^3/day were measured, resulting in a cumulative gas output of approximately 13,000 Sm^3. However, it has been indicated that rates larger than 5,00,000 Sm^3/day may be needed at current rates for the production to be economically viable [32–35].

7.1.4 Role of Modeling and Simulation

The assessment of the production potential of CH_4 from hydrates involves modeling a complex system consisting a variety of coupled processes. The reliability of these predictions depends on: (a) robustness of the mathematical model, (b) accurate knowledge of relevant parameters, and (c) availability of experimental results for model validation. Such complex and nonlinear mathematical models typically defy analytical solutions making numerical simulation to be central to the assessment of the gas production from hydrates. Numerical simulation, furthermore, allows exploration of alternative scenarios that would be otherwise prohibitively expensive [36]. Finally, considering the parametric uncertainties, numerical simulations provide us envelopes of possibilities leading to the identification of key parameters.

7.2 Mathematical Modeling

Equations governing flow of gases and liquids in porous media including heat transfer, phase change, and chemical reactions are set up in the following sections. A single phase and a more elaborate multiphase model are described. Results arising from the simulation of these models and their validation are discussed. The goal of simulation remains the amount of methane recovered and the quantity of CO_2 retained in the reservoir in hydrate form.

7.2.1 Single-Phase Model

The mathematical model described here is derived from the work of Ahmadi et al. [28, 29] with a few changes. The important difference is that their work did not account for variation of density with respect to temperature and pressure as a part of their governing equations. This study uses the ideal gas state equation for methane gas as a function of pressure and temperature thus entirely eliminating the variable density term. An underground reservoir that is partially saturated with CH_4 hydrate and also contains free CH_4 gas at the reservoir pressure P_e and reservoir temperature T_e is considered for this analysis. Initially, the hydrates in the reservoir are assumed to be in a state of thermodynamic equilibrium with their surroundings. Upon drilling of a production well, the pressure in the well bore is maintained at a constant value P_{dep}, which is far below the hydrate equilibrium pressure P_{eq} at temperature $T_{eq} = T_e$. Such conditions initiate hydrate dissociation at the well region thus leading to formation of gas and water. The process of hydrate dissociation then pervades radially outward from the well with time. It is assumed that the hydrate dissociation occurs in a narrow region, which can be treated as the dissociation front. Figure 7.4 shows a schematic drawing of the axisymmetric reservoir.

The dissociation front divides the reservoir into two zones. The near-well 'Gas Zone' (denoted by index $n = 1$) contains CH_4 gas and water, while the far-well 'Hydrate Zone' (denoted by index $n = 2$) on beyond the dissociation front contains the undissociated CH_4 hydrates with free CH_4 gas. An inherent assumption in the model is that the temperature and pressure distributions are axisymmetric with the well bore acting as the central axis. The modeling of CH_4 gas and energy flow in this study involves several assumptions as stated below:

Fig. 7.4 Schematic of the axisymmetric reservoir used in single-phase calculation

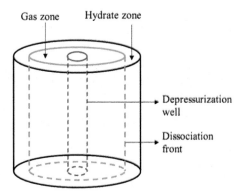

1. Mass and momentum conservation equations of the water phase are ignored. Hence, the mathematical model is developed for a single phase. In the absence of a gas–liquid interface, capillary pressure effects have not been considered in this study.
2. At the scale of the REV (Chap. 1), various components of the system such as CH_4 gas, water, CH_4 hydrate, and the porous rock medium are in thermal equilibrium with each other in their respective zones and have the same temperature at any given location in space and time. However, the reservoir itself is not isothermal.
3. The pressure and temperature at the dissociation front are the equilibrium pressure, P_{eq}, and temperature, T_{eq}, but the front moves progressively outwards with time. The phase boundary connecting variables P_{eq} and T_{eq} obey the Makogon correlation [24].
4. Effective medium properties such as porosity (ϕ), permeability in the two zones (K_{1abs} and K_{2abs}), and equivalent thermal conductivities in the two zones (K_{1eq} and K_{2eq}) are constant with respect to time and location.
5. Methane gas obeys the ideal gas state equation; in view of large changes in pressure, the Joule–Thomson cooling effect for gas is included in the energy conservation equation. Gas saturations in the two zones (s_{1g} and s_{2g}) do not vary with time or location.
6. Effects of phenomena such as mass diffusion, heat dispersion due to porous formation, solubility of gas in water, and the presence of water vapor in gas phase are not accounted for.

7.2.2 Governing Equations

The mass conservation equation for gas flow in porous media in a radially axisymmetric coordinate system cylindrical polar coordinates is given as:

$$\frac{\partial(\phi s_g \rho_g)}{\partial t} + \frac{1}{r}\frac{\partial(r\rho_g v_{gr})}{\partial r} = 0 \qquad (7.1)$$

Here ρ_g is gas density and v_{gr} is the radial bulk velocity in porous media. The bulk velocity in this analysis obeys Darcy's law for flow in porous media, given as:

$$v_r = -\frac{K_{abs}}{\mu_g}\frac{\partial(P_g)}{\partial r} \qquad (7.2)$$

Here K_{abs} is the absolute permeability of the porous medium, μ_g is gas viscosity, and P_g is gas pressure. Equation 7.2 does not include relative permeability because capillary effects are neglected. The ideal gas state equation for CH_4 gas is given as:

$$\rho_g = \frac{P_g M^m}{ZRT} \qquad (7.3)$$

Here M^m is the molar mass of CH_4 gas, Z is the gas compressibility factor, R is the universal gas constant, and T is the temperature. Bulk velocity from Eq. 7.2 and gas density from Eq. 7.3 are substituted in the mass conservation Eq. 7.1 which leads to a pressure equation as follows [30]:

$$\frac{\partial(P_g)}{\partial t} - \frac{K_{abs}}{2\phi s_g \mu_g}\left[\frac{\partial^2(P_g^2)}{\partial r^2} + \frac{1}{r}\frac{\partial(P_g^2)}{\partial r}\right] = \frac{P_g}{T}\left(\frac{\partial T}{\partial t} - \frac{K_{abs}}{\phi s_g \mu_g}\frac{\partial P_g}{\partial r}\frac{\partial T}{\partial r}\right) \qquad (7.4)$$

The final form of the energy equation may be derived in a similar manner by using the expressions for density and bulk velocity in the energy conservation equation in porous medium and is given as:

$$\frac{\partial}{\partial t}\left[\left(\sum_{\gamma=l,\,g,\,h}(\phi s_\gamma \rho_\gamma C_{p\gamma}) + (1-\phi)\rho_s C_{ps}\right)T\right] + \frac{1}{r}\frac{\partial(r\rho_g v_{gr}T)}{\partial r}$$
$$= \frac{1}{r}\frac{\partial}{\partial r}\left(rK_{eq}\frac{\partial T}{\partial r}\right) - \sigma_g\left[\phi s_g \rho_g\frac{\partial P_g}{\partial t} + \frac{\rho_g v_{gr}}{r}\frac{\partial(rP_g)}{\partial r}\right] \qquad (7.5)$$

Here $C_{p\gamma}$ is the heat capacity for the phase γ, σ_g is the coefficient for Joule–Thomson cooling effect, and K_{eq} is the equivalent thermal conductivity. At the dissociation front, CH_4 gas mass balance and energy flux balance have to be enforced to determine the existing equilibrium P-T conditions. The mass balance of gaseous CH_4 at the dissociation front may be written as [31]:

$$\rho_g v_{1gr} = \rho_g v_{2gr} + \phi\frac{\partial R}{\partial t}\left[\rho_g(s_{2g} - s_{1g}) + \frac{M^m s_h \rho_h}{(M^m + N_h M^w)}\right] \qquad (7.6)$$

Here v_{1gr}, v_{2gr} represent gas velocities at the dissociation front, s_{1g}, s_{2g} represent the gas saturations in zones 1 and 2, respectively. The derivative $\frac{\partial R}{\partial t}$ is the speed of the dissociation front which keeps moving radially outwards with time, ρ_g is the gas density at the dissociation front and it assumes a unique value that is evaluated using the ideal gas law.

Similarly, energy balance at the dissociation front is given by:

$$\rho_g v_{1gr} T_{eq} C_{pg} - K_{1eq} \left.\frac{\partial T_1}{\partial r}\right|_{r=R} = \rho_g v_{2gr} T_{eq} C_{pg} - K_{2eq} \left.\frac{\partial T_2}{\partial r}\right|_{r=R} + \frac{\Delta H \dot{m}_h}{2\pi R} \qquad (7.7)$$

Here, K_{1eq} and K_{1eq} represent the equivalent thermal conductivities in zones 1 and 2, respectively. ΔH is the heat of CH_4 hydrate dissociation. $\dot{m}_h$ is the mass rate of dissociation of CH_4 hydrate and can be evaluated as a function of the speed of dissociation front as follows [32]:

$$\dot{m}_h = 2\pi R \rho_h s_h \frac{\partial R}{\partial t} \qquad (7.8)$$

The saturations for various components of the system obey the relations $s_{1g} + s_w = 1$ for zone 1 and $s_{2g} + s_h = 1$ for zone 2. Given the fact that initially there is no water in the reservoir and movement of water has been ignored in this study, the water saturation in zone 1 can be calculated using the conservation relation as [33]:

$$s_w = s_h \frac{N_h M^w}{M^m + N_h M^w} \frac{\rho_h}{\rho_w} \qquad (7.9)$$

7.2.3 Constitutive Relations

Along with the governing equations for mass, momentum, and energy transfer, a suite of constitutive relations are necessary to complete the model for a solution. The equilibrium pressure–temperature correlation for this study has been taken from Makogon's work as [24]:

$$\log_{10}(P_{eq}) = (0.0342 \times (T_{eq} - T_o)) + (0.0005 \times (T_{eq} - T_o)^2) + 6.4804 \qquad (7.10)$$

Here P_{eq} is in Pascal and T_{eq} is in Kelvin. The process of hydrate dissociation at the dissociation front is an endothermic phase change process. The dissociation heat (J/Kg) of hydrate has been given by Kamath [] as:

$$\Delta H = 1050 \times T_{eq} + 3527000 \tag{7.11}$$

The equivalent thermal conductivity in this analysis has been calculated assuming a parallel mode of conduction and is given as:

$$\text{zone 1: } K_{1eq} = \phi s_w K_w + \phi s_{1g} K_g + (1 - \phi) K_{sw}$$
$$\text{zone 2: } K_{2eq} = \phi s_h K_h + \phi s_{2g} K_g + (1 - \phi) K_{sd} \tag{7.12}$$

Series mode of heat conduction is also possible in the reservoir, and the equivalent thermal conductivity is given as [34]:

$$\text{zone 1: } \frac{1}{K_{1eq}^s} = \frac{\phi s_w}{K_w} + \frac{\phi s_{1g}}{K_g} + \frac{(1 - \phi)}{K_{sw}}$$
$$\text{zone 2: } \frac{1}{K_{2eq}^s} = \frac{\phi s_h}{K_h} + \frac{\phi s_{2g}}{K_g} + \frac{(1 - \phi)}{K_{sd}} \tag{7.13}$$

7.2.4 Initial and Boundary Conditions

The well-bore center is at $r = 0$, and the radius of the well is taken as $r_o = 20$ mm. Symbol $R(t)$ is the location of the dissociation front from the center of the well. In the present study, it is assumed that the far end of the reservoir, which is initially at a distance $L = 100$ m from the center in the reservoir, shifts radially outward, like the dissociation front. Therefore, the far end of the reservoir is always at a radius $r = R(t) + L$. This means that the computational domain ranges from $r = r_o$ to $r = R(t) + L$. However, simulations using the single-phase mathematical model have been performed separately for the two zones. The P-T boundary conditions at the dissociation front separating the two zones are ultimately determined by the mass and energy balance fluxes which are enforced at this location.

The depressurization pressure at the well bore is kept constant at P_{dep} at $r = r_o$. Ideally, the far end of the reservoir should be at an infinite distance with unchanging pressure and temperature conditions, but that is not computationally possible. Hence, for the far end of the reservoir located at $r = R(t) + L$, it is assumed that the reservoir pressure and temperature are fixed at P_e and T_e. The following initial and boundary conditions have been used for this analysis.

$$P_g(r, 0) = P_e; P_g(r_o, t) = P_{dep}; P_g[R(t), t] = P_{eq}; P_g[R(t) + L, t] = P_e$$
$$T(r, 0) = T_e; \frac{\partial T(r_o, t)}{\partial r} = 0; T[R(t), t] = T_{eq}; T[R(t) + L, t] = T_e \tag{7.14}$$

Results obtained from the single-phase model of methane extraction are discussed in Sect. 7.4.

7.3   Two-Phase Model

In this section, we present a general multispecies multiphase kinetic model developed for simulation of hydrate decomposition and formation in a geological reservoir. The species accounted for are water (H_2O), methane (CH_4), and carbon dioxide (CO_2). Transport is considered over three mobile phases (i.e., aqueous ($\gamma = l$), gas ($\gamma = g$), and liquid carbon dioxide ($\gamma = n$)) and for immobile phases (i.e., CH_4 hydrate ($\gamma = mh$), CO_2 hydrate ($\gamma = ch$), ice ($\gamma = i$), and geologic media ($\gamma = s$)). It should be noted that gas hydrates are considered as a phase of the two species (either H_2O and CH_4, or H_2O and CO_2) and are not independently referred. However, unlike previous studies by McGrail et al. [35], CH_4 and CO_2 gas hydrates are considered two different phases in this study. It should also be noted that a phase is a physical manifestation of (and is composed of) one or more species. For example, H_2O may be found in liquid (water), gaseous (vapor), and solid (ice and hydrate) phases. Therefore, physical properties in this model are assigned to a species in terms of the phases in which that species may be found. For an initial discussion on the subject, see Sect. 2.6.2 of Chap. 2.

7.3.1   Governing Equations

The mass, momentum, and energy conservation equations as well as the kinetic relations are given below [36, 37]:

$$\frac{\partial}{\partial t}\left[\sum_{\gamma=l,\,n,\,g,\,h,\,i,\,p}\left(\phi_D\omega_\gamma^i\rho_\gamma s_\gamma\right)\right] = -\sum_{\gamma=l,\,n,\,g}\nabla\left(\omega_\gamma^i h_\gamma \mathbf{V}_\gamma\right)$$

$$-\sum_{\gamma=w,\,a,\,o}\nabla\left(\mathbf{J}_\gamma^i\right) + \sum_{\gamma=l,n,g}\nabla\left(\omega_\gamma^i \dot{m}_\gamma\right) - \dot{m}_h^i \quad \text{for} \quad i = w,\,m,\,c,\,s \qquad (7.15)$$

where $\mathbf{J}_\gamma^i = -\phi_D\rho_\gamma s_\gamma \frac{M^i}{M_\gamma}\left(\tau_\gamma D_\gamma^i + \mathbf{D}_{h_\gamma}\right)\nabla\chi_\gamma^i$

$\dot{m}_h^i = \frac{\partial}{\partial t}\left(\phi_D\omega_h^i\rho_h s_h\right)$ for $i = m,\,c$; $\dot{m}_h^i = 0$ for $i = w$

$$\mathbf{V}_\gamma = -\frac{k_{r\gamma}\mathbf{k}}{\mu_\gamma}\left(\nabla P_\gamma + \rho_\gamma g\mathbf{z}_g\right) \qquad (7.16)$$

$$\frac{\partial}{\partial t}\left[\sum_{\gamma=l,\,n,\,g,\,h,\,i,p}\left(\phi_D\rho_\gamma s_\gamma U_\gamma\right)+(1-\phi_T)\rho_s s_s U_s+(\phi_T-\phi_D)\rho_l U_l\right]-\dot{q}=$$

$$-\sum_{\gamma=l,\,n,\,g}\nabla\left(\rho_\gamma h_\gamma \mathbf{V}_\gamma\right)-\sum_{i=w,\,m,\,c,\,s}\nabla\left(h_g^i\mathbf{J}_g^i\right)\quad(7.17)$$

$$+\nabla(\mathbf{k}_e\nabla T)+\sum_{\gamma=l,\,n,\,g}\nabla\left(h_\gamma\dot{m}_\gamma\right)$$

Formation/dissociation:

$$Gas.n_g^{h,\,i}H_2O\rightleftharpoons Gas+n_g^{h,\,i}H_2O\ \text{ for }\ i=c,m$$

$$\frac{d[C_h]}{dt}=k_r(T)\left(P_g-P_e\right)\tag{7.18}$$

Formation:

$$[\dot{m}_{ih}]^f=M_{ih}\left(\phi A_{SH}s_l+\phi^2 A_{SH}s_l s_h\right)K_f^i\exp\left(-\frac{E^i}{RT}\right)\left(P_g^i-P_{eq}^i\right)$$

Dissociation:

$$[\dot{m}_{ih}]^d=M_{ih}\left(\phi^2 A_{SH}s_l s_h\right)K_d^i\exp\left(-\frac{E^i}{RT}\right)\left(P_{eq}^i-P_g^i\right)$$

Mass exchange:

$$\dot{m}_h^i=K_h\left(P_g^i-\varphi_g^i P_h^{eq}\right)\ \text{ for }\ i=c,\ m;\ \dot{m}_h^i=0\ \text{ for }\ i=w,\ s$$

We have four governing equations (three for mass conservation and one for energy). Each governing equation is solved for a single variable that is referred to as the primary unknown. However, each of these equations involves numerous variables such as density, enthalpy along with a set of coefficients within the fundamental equations that are dependent upon and linked to the primary variables. These are known as the secondary variables. The choice of primary and secondary variables is, in principle, not subject to any rules. Constitutive relations are required for providing the functional links between the primary and secondary variables of the governing equations [38, 39]. These relations also help in completely specifying the thermodynamic and hydrologic state within a porous media system given a sufficient number of independent intensive property values. However, a dearth of such relationships and intensive property values necessitates simplification of the present model to make it mathematically determinable.

In the following discussions, we develop a non-isothermal, multiphase, multispecies model for simulation of CH_4 gas production via depressurization and simultaneous CO_2 sequestration using the primary governing equations presented above. A one-dimensional Cartesian reservoir with a production and an injection well at its ends is represented. Figure 7.5 shows the schematic of the reservoir considered in this study. The reservoir is core (of length L and cross-sectional area A_c) which contains a hydrate-bearing layer that is sandwiched between

Fig. 7.5 Schematic of the 1-D Cartesian reservoir

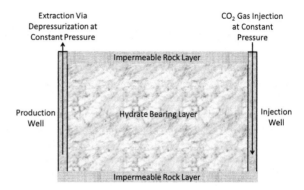

impermeable rock sediment layers. It is initially partially saturated with only CH_4 hydrates and also contains water and free CH_4 gas. The CH_4 gas inside the reservoir is in equilibrium with the hydrates at a given initial reservoir temperature T_{init} and reservoir pressure P_{init}. The production well (on the left) is depressurized at a constant bottom-hole pressure P_{dep} and simultaneously, CO_2 is injected at a constant pressure P_{inj}^c (below its liquefaction pressure) at the injection well (on the right).

The following assumptions have been made for simplification of the governing differential equations of the model as follows [40–42]:

1. With REV, all phases present in the system are in thermal equilibrium with each other and have the same temperature at any given location in space and time. However, the reservoir itself is not isothermal.
2. The mode of hydrate dissociation is kinetic. The model allows for hydrate dissociation as well as formation. There is no sharp dissociation front, and dissociation can occur anywhere depending on the local *P-T* conditions.
3. Mass diffusion and dispersion fluxes have been ignored in the study. Subsequently, heat fluxes due to momentum diffusion and dispersion have also been neglected.
4. CH_4 gas and CO_2 gas obey the ideal gas law. Joule–Thomson cooling effect for during gas flow in pores is ignored in this study. Compressibility factor for calculating density is set as $Z = 1$.
5. The total and diffusive porosities are the same, and there is no thermal capacitance term in the energy equation. This implies $\phi_T = \phi_D = \phi$.
6. Solubility of CH_4 and CO_2 gases in the aqueous phase and the presence of water vapor in gas phase are assumed to be low and are ignored in the study. This implies that the gas phase comprises CH_4 and CO_2 gas, and aqueous phase is made up of pure water. Mathematically, this implies $\omega_g^m + \omega_g^c = 1$ and $\omega_l^w = 1$.
7. The pressure–temperature conditions inside the reservoir are always such that water is in liquid state, and ice formation does not take place. Hence, ice phase is not taken into account.

8. The pressure–temperature conditions inside the reservoir during operation are such that CO_2 remains in the gaseous phase. Hence, the non-aqueous liquid phase for CO_2 is not taken into account.

9. Aqueous phase density as well as CH_4 and CO_2 hydrate densities is invariant with respect to pressure or temperature. Hydration number for both hydrates is constant at $N_h = 6$.

10. In reality, the equilibrium conditions for hydrate dissociation/formation are different in case of a binary mixture of two hydrate-forming gases. For modeling kinetics of hydrate dissociation/formation of the hydrates, separate equilibrium data for pure CH_4 and CO_2 hydrates have been considered in the present work.

11. Direct molecular exchange of CH_4 and CO_2 gas molecules in hydrates has not been accounted for.

12. Water soluble salt precipitates have been ignored in this study. Effect of gravity is not included as it facilitates 1-D modeling of transport in the horizontal plane.

7.3.2 Equilibrium Data for Hydrate Stability

Equilibrium data for hydrates is central to any calculations for rates of formation/dissociation of hydrates. Three-phase equilibrium data (liquid, hydrate, and vapor) of water rich of pure CH_4 and CO_2 hydrates is taken from experimental database [43]. The tabulated equilibrium P-T data values from this work are fitted to polynomial curves as a function of temperature, and the equations obtained are given as follows [44–47]:

$$P_{eq}^m = 0.1588 \left[\frac{(T - 280.6)}{4.447} \right]^3 + 0.6901 \left[\frac{(T - 280.6)}{4.447} \right]^2$$
$$+ 2.473 \left[\frac{(T - 280.6)}{4.447} \right] + 5.513 \text{ in MPa} \tag{7.19}$$

$$P_{eq}^c = 0.06539 \left[\frac{(T - 278.9)}{3.057} \right]^3 + 0.2738 \left[\frac{(T - 278.9)}{3.057} \right]^2$$
$$+ 0.9697 \left[\frac{(T - 278.9)}{3.057} \right] + 2.479 \text{ in MPa} \tag{7.20}$$

7.3.3 Porosity and Absolute Permeability

As the saturation of hydrates change in the reservoir, the effective porosity for fluid flow changes because of changing pore space and influences absolute permeability of a porous medium. The effective fluid porosity, ϕ_{lg}, is defined as [48]:

$$\phi_{lg} = (1 - s_h)\phi \tag{7.21}$$

An empirical relationship between ϕ_{lg} and K_{abs} for Berea sandstone is given by [49]:

$$\begin{aligned}
\phi_{lg} \leq 0.11 &: K_{abs} = 5.51721(\phi_{lg})^{0.86} \times 10^{-15} \text{ m}^2 \\
\phi_{lg} > 0.11 &: K_{abs} = 4.84653(\phi_{lg})^{0.86} \times 10^{-15} \text{ m}^2
\end{aligned} \tag{7.22}$$

This model has been adopted in the present work.

7.3.4   Relative Permeability and Capillary Pressure

The relative permeability of the respective phases in the present model is evaluated using the Brooks–Corey function specified as [50]:

$$\text{liquid phase: } k_{rl} = \left(\frac{\frac{s_l}{s_l + s_g} - s_{lr}}{1 - s_{lr} - s_{gr}} \right)^{n_l} \tag{7.23}$$

$$\text{gas phase: } k_{rg} = \left(\frac{\frac{s_g}{s_l + s_g} - s_{gr}}{1 - s_{lr} - s_{gr}} \right)^{n_g} \tag{7.24}$$

The pressures of the gaseous and aqueous phases differ by an amount called the *capillary pressure* and are obtained as:

$$P_c = P_g - P_w = P_c = P_{ec} \left(\frac{\frac{s_l}{s_l + s_g} - s_{lr}}{1 - s_{lr} - s_{gr}} \right)^{-n_c} \tag{7.25}$$

7.3.5   Gas Phase Viscosity

The gas phase viscosity is computed by combining pure component vapor viscosities, according to the mixing rules of Wilke [51]:

$$\mu_g = \frac{\chi_g^m \mu_g^m}{\chi_g^m + \chi_g^c \varphi_g^{cm}} + \frac{\chi_g^c \mu_g^c}{\chi_g^c + \chi_g^m \varphi_g^{mc}} \; ; \; \varphi_g^{ij} = \frac{\left[1 + \left(\frac{\mu_g^i}{\mu_g^j} \right)^{1/2} \left(\frac{M^j}{M^i} \right)^{1/4} \right]^2}{\left[8 \left(1 + \frac{M^i}{M^j} \right) \right]^{1/2}} \tag{7.26}$$

7.3.6 Specific Heat Capacities

The specific heat capacities for the two hydrate phases and solid rock medium are taken invariant with respect to pressure or temperature. The specific heat capacities for CH_4 (gas) and water (liquid) are taken from available experimental results [51]. The specific heat capacity for CO_2 (gas) is taken from other sources. The correlations for CH_4, CO_2, H_2O, respectively, in J/Kg-K are given by

$$C_{pg}^m = 1238.79 + (3.1303T) + (7.905 \times 10^{-4}T^2) - (6.858 \times 10^{-7}T^3)$$
$$C_{pg}^c = 505.11 + (1.1411T) - (89.139 \times 10^{-5}T^2) + (210.566 \times 10^{-9}T^3) \quad (7.27)$$
$$C_{pl}^w = 4023.976 + (.57736T) - (8.314 \times 10^{-5}T^2)$$

7.3.7 Heat of Hydrate Formation

Enthalpies of the reactions arise for gas dissociation into gas and either water or ice. These are determined directly by analysis from the Clapeyron equation. The tabulated data for enthalpy of reactions (J/Kg-K) is fitted with a polynomial as a function of temperature. The equations obtained for CH_4- and CO_2-hydrates, respectively, are obtained as follows [52]:

$$
\begin{aligned}
\Delta H_{mh}^f(T) = &\left[30.1\left(\frac{T-296.0}{14.42}\right)^9\right] - \left[12.94\left(\frac{T-296.0}{14.42}\right)^8\right] \\
&- \left[160.1\left(\frac{T-296.0}{14.42}\right)^7\right] + \left[69.12\left(\frac{T-296.0}{14.42}\right)^6\right] \\
&+ \left[285.8\left(\frac{T-296.0}{14.42}\right)^5\right] - \left[119.2\left(\frac{T-296.0}{14.42}\right)^4\right] \\
&- \left[193.9\left(\frac{T-296.0}{14.42}\right)^3\right] + \left[68.22\left(\frac{T-296.0}{14.42}\right)^2\right] \\
&+ \left[37.07\left(\frac{T-296.0}{14.42}\right)\right] + 420.1
\end{aligned}
\quad (7.28)
$$

$$\Delta H_{ch}^{f}(T) = \left[2.528\left(\frac{T - 278.15}{2.739}\right)^{8}\right] + \left[.07536\left(\frac{T - 278.15}{2.739}\right)^{7}\right]$$
$$- \left[9.727\left(\frac{T - 278.15}{2.739}\right)^{6}\right] + \left[1.125\left(\frac{T - 278.15}{2.739}\right)^{5}\right]$$
$$+ \left[4\left(\frac{T - 278.15}{2.739}\right)^{4}\right] - \left[4.154\left(\frac{T - 278.15}{2.739}\right)^{3}\right]$$
$$+ \left[14.43\left(\frac{T - 278.15}{2.739}\right)^{2}\right] - \left[6.668\left(\frac{T - 278.15}{2.739}\right)\right] + 389.9$$

$$(7.29)$$

7.3.8 Equivalent Thermal Conductivity

Series and parallel modes of thermal conductivities are calculated as follows [53]:

$$K_{eq} = \phi s_{l} K_{l} + \phi s_{g} K_{g}^{m} + \phi s_{g} K_{g}^{c} + \phi s_{mh} K_{h}^{m} + \phi s_{ch} K_{h}^{c} + (1 - \phi) K_{s}$$
$$\frac{1}{K_{eq}^{s}} = \frac{\phi s_{l}}{K_{l}} + \frac{\phi s_{g}}{K_{g}^{m}} + \frac{\phi s_{g}}{K_{g}^{c}} + \frac{\phi s_{mh}}{K_{mh}} + \frac{\phi s_{ch}}{K_{ch}} + \frac{(1 - \phi)}{K_{s}} \qquad (7.30)$$

7.3.9 Initial and Boundary Conditions

The simulations have been carried out for a one-dimensional $(L \times A_{cs})$ reservoir with the following two boundaries: (1) production well (left) and (2) injection well (right). The length of the reservoir is taken as $L = 100$ m for the baseline simulation. The solution for all variables is calculated along the x-direction. Initially, the reservoir is partially saturated with CH_4 hydrates and contains free CH_4 gas and water. The reservoir is at temperature T_{init}, and the CH_4 gas pressure is the equilibrium pressure for CH_4 hydrates at this temperature given by P_{init}. The depressurization pressure at the production well $(x = 0)$ is kept constant at P_{dep}. CO_2 gas is injected at a constant partial pressure of P_{inj}^c at the injection well. If we consider that the injection well is placed in a reservoir with a network of several other production and injection wells distributed symmetrically around it, then the partial pressure of CH_4 gas forms the symmetry boundary condition at the injection well. A similar argument can be made for the temperature conditions at both the production and injection wells. However, a constant temperature T_{inj} (Dirichlet boundary condition) was used for temperature at the injection well assuming that CO_2 can be injected at elevated temperatures so as to maintain the injection well temperature. Accordingly, for the present work, the following initial and boundary conditions are applicable:

$$P_g(x, 0) = P_{init}; \ P_g(0, t) = P_{dep}; \ P_c(0, t) = P_{ec}; \ P_g^c(L, t) = P_{inj}^c;$$

$$\frac{\partial P_g^m(L, t)}{\partial x} = 0; \ P_g(L, t) = P_g^m(L, t) + P_g^c(L, t)$$

$$T(x, 0) = T_{init}; \ \frac{\partial T_{dep}(0, t)}{\partial x} = 0; \ \frac{\partial T_{inj}(L, t)}{\partial x} = 0$$

$$\omega_g^m(x, 0) = 1; \ \frac{\partial \omega_g^m(0, t)}{\partial x} = 0; \ \frac{\partial \omega_g^c(0, t)}{\partial x} = 0 \qquad (7.31)$$

$$s_g(x, 0) = s_{og}; \ \frac{\partial s_g(0, t)}{\partial x} = 0; \ \frac{\partial s_g(L, t)}{\partial x} = 0$$

$$s_l(x, 0) = s_{ol}; \ \frac{\partial s_l(L, t)}{\partial x} = 0; \ s_h(x, 0) = s_{mh}(x, 0) = s_{omh}$$

7.4   Results and Discussion

The single- and multiphase models developed in the previous section have been numerically solved and extensively validated against published results. Representative results are described in the discussions below, Sects. 7.4.1–7.4.2 for the single-phase model and 7.4.3–7.4.6 for the multiphase. Tables 7.1 and 7.2 carry model parameters employed in the present simulation. Detailed discussion on the range of parameters that can be treated as realistic may be found elsewhere [54, 55].

7.4.1   Evolution of Pressure and Temperature Profiles

Figure 7.6 show the evolution of the pressure and temperature in a CH_4 hydrate reservoir. The results are presented for time instants of 10 and 40 days. Figure 7.6a shows the evolution of pressure in the reservoir. The reservoir is initially assumed to be at stable conditions of 15 MPa and 287 K. The production well pressure is maintained at a constant value of 2 MPa. A sharp change in CH_4 gas pressure after 10 days of operation can be seen, which is coincident with the dissociation front. The location of this sharp change shifts radially outwards with time, as can be observed in the pressure profile at 40 days. Similarly, a sharp change at the same location can be seen in the temperature distribution in Fig. 7.6b. These results are compared with two previous studies based on a similar axisymmetric equilibrium model: (1) Ahmadi et al. [28] and (2) Ji et al. [27], as shown in Fig. 7.6c (for pressure) and Fig. 7.6d (for temperature). Ji et al. [27] presented a linearized solution for pressure- and temperature-based Makogon's approach [24]. Though Ahmadi et al. [28] improved significantly over the work of [27], momentum-energy coupling and heat conduction in the gas zone were ignored in their study.

Table 7.1 Parameters used in single-phase simulation

Symbols	Physical meaning	Value
ϕ	Porosity	0.2
μ_g	Viscosity of gas	1×10^{-5} Pa-s
K_{1abs}	Absolute permeability of gas zone	8×10^{-15} m^2
K_{2abs}	Absolute permeability of hydrate zone	1×10^{-15} m^2
s_w	Saturation of water in gas zone	0.15
s_h	Saturation of CH$_4$ hydrate in hydrate zone	0.19
Z	Gas compressibility factor	1.0
M^w	Molar mass of water	0.018015 kg/mol
M^m	Molar mass of methane	0.016042 kg/mol
N_h	Hydration number	6.0
K_w	Water conductivity	0.59 W/m-K
K_h	CH$_4$ hydrate conductivity	0.62 W/m-K
K_g	CH$_4$ gas conductivity	0.03 W/m-K
K_{sd}	Solid rock medium conductivity (dry)	3.4 W/m-K
K_{sw}	Solid rock medium conductivity (wet)	0.5 W/m-K
C_{pw}	Water specific heat capacity	4181 J/kg-K
C_{ph}	CH$_4$ hydrate specific heat capacity	2150 J/kg-K
C_{pg}	CH$_4$ gas specific heat capacity	2206 J/kg-K
C_{ps}	Solid rock medium specific heat capacity	920 J/kg-K
ρ_w	Water density	1000 kg/m^3
ρ_h	Hydrate density	930 kg/m^3
ρ_{og}	CH$_4$ gas density at standard conditions	0.714 kg/m^3
ρ_s	Rock density	2675 kg/m^3
σ_g	Throttling coefficient	-1.5×10^{-4} J/kg-Pa
T_o	Standard temperature	273.15 K
P_o	Standard pressure	101,325 Pa
P_{dep}	Depressurization well pressure	2 MPa
P_e	Initial reservoir pressure	15 MPa
T_e	Initial reservoir temperature	287 K

As clearly seen in Fig. 7.6, the present model predicts a much slower movement of the dissociation front into the reservoir. This has direct implications on the gas production. As discussed in previous chapters, the dissociation front separates the reservoir into two zones, where the near-well gas zone is devoid of any hydrate. The gas zone keeps expanding into the reservoir and the faster it does so, the higher is the production. Hence, the models proposed by [27, 28] grossly overpredict CH$_4$ gas production from the reservoir. The present results show that momentum-energy coupling leads to a nearly isothermal temperature distribution in the respective zones, unlike the results presented in the previous simplistic models. Also, the hydrate equilibrium conditions of P-T at the decomposition front are found to be

Table 7.2 Parameters used in two-phase simulation

Symbol	Physical meaning	Value
ϕ	Porosity	0.28
C_{pmh}	Specific heat capacity for CH_4 hydrate phase	2220 J/kg-K
C_{pch}	Specific heat capacity for CO_2 hydrate phase	2220 J/kg-K
C_{ps}	Solid rock phase specific heat capacity	835 J/kg-K
K_l	Thermal conductivity of H_2O in aqueous phase	0.59 W/m-K
K_g^m	Thermal conductivity of CH_4 in gaseous phase	0.030 W/m-K
K_g^c	Thermal conductivity of CH_4 in gaseous phase	0.015 W/m-K
K_{mh}	Thermal conductivity of CH_4 hydrate phase	0.62 W/m-K
K_{ch}	Thermal conductivity of CO_2 hydrate phase	0.62 W/m-K
K_s	Thermal conductivity of solid rock phase	3.5 W/m-K
s_{lr}	Irreducible aqueous phase saturation	0.2
s_{gr}	Irreducible gas phase saturation	0.0
s_{ol}	Initial aqueous phase saturation	0.3
s_{og}	Initial gas phase saturation	0.1
s_{omh}	Initial CH_4 hydrate saturation	0.6
s_{och}	Initial CO_2 hydrate saturation	0.0
R	Universal gas constant	8.314472 J/Kmol
M^w	Molar mass of H_2O	0.018015 kg/mol
M^m	Molar mass of CH_4	0.016042 kg/mol
M^c	Molar mass of CO_2	0.044010 kg/mol
N_h	Hydration number	6.0
ρ_l	Aqueous phase density	1000 kg/m^3
ρ_{mh}	CH_4 hydrate phase density	919.7 kg/m^3
ρ_{ch}	CO_2 hydrate phase density	1100 kg/m^3
ρ_s	Solid rock phase density	2675 kg/m^3
K_f^m	Intrinsic rate constant for CH_4 hydrate formation	0.00290 mol/Pa-s-m^2
K_d^m	Intrinsic rate constant for CH_4 hydrate dissociation	123.96 kmol/Pa-s-m^2
K_f^c	Intrinsic rate constant for CO_2 hydrate formation	0.00035 mol/Pa-s-m^2
K_d^c	Intrinsic rate constant for CO_2 hydrate dissociation	123.96 kmol/Pa-s-m^2
E^m	Activation energy for CH_4 hydrate	81.084 kJ/mol
E^c	Activation energy for CO_2 hydrate	81.084 kJ/mol
A_{SH}	Specific area of hydrates	3,75,000 m^2/m^3
μ_l	Aqueous phase viscosity	0.001 Pa-s
μ_g^m	Viscosity of CH_4 in gas phase	1.35×10^{-5} Pa-s
μ_g^c	Viscosity of CO_2 in gas phase	1.48×10^{-5} Pa-s
P_{ec}	Entry capillary pressure	5000 Pa
n_l	Constant for aqueous phase relative permeability	4
n_g	Constant for gas phase relative permeability	2
n_c	Constant for capillary pressure	0.65

(continued)

Table 7.2 (continued)

Symbol	Physical meaning	Value
T_{init}	Initial reservoir temperature	283 K
P_{init}	Initial CH_4 gas partial pressure	7.07 MPa
P_{inj}^c	CO_2 gas injection partial pressure	4.5 MPa
P_{dep}	Depressurization well pressure	3.5 MPa
Λ	Constant for longitudinal heat transfer	1 W/m²-K
P_{cs}	Perimeter of the core cross-section	40 m
A_{CS}	Area of the core cross-section	100 m²

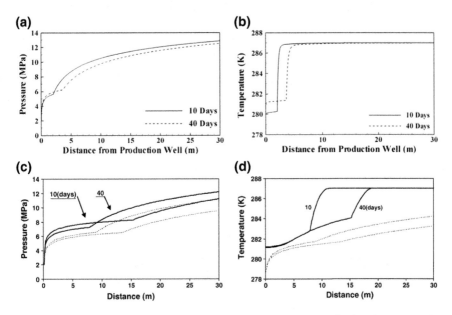

Fig. 7.6 **a** Pressure distribution (present work), **b** temperature distribution (present work), **c** pressure distribution, (solid lines are from [28] and dashed lines are from [27]; **d** temperature distribution, (solid lines denoting Ahmadi et al. [28] and dashed lines are from Ji et al. [27]). The comparisons are on the same scale

much lower in this study. This result indicates that the *P-T* conditions are influenced more by the production well pressure than the reservoir initial conditions.

Figure 7.7 shows the variation of the speed of the decomposition front with respect to its location in the reservoir over a span of 40 days. Initial location of the front is assumed to be at 1 m from the production well. The speed of the front rapidly decreases as the dissociation front moves deeper into the reservoir. This indicates that the production of CH_4 gas will gradually fall. The small serrations seen in the plots are numerical in origin and occur due to the discontinuities induced by grid shifting for the decomposition front. These are absent in the multiphase model.

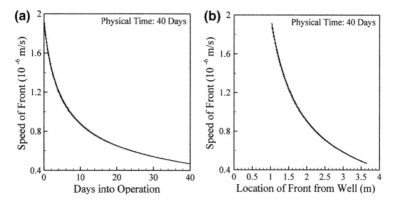

Fig. 7.7 a Variation of speed of decomposition front with time; **b** variation of speed of decomposition front with location of the front in the reservoir

7.4.2   Sensitivity Analysis

Gas production from hydrate reservoirs is a sensitive function of several thermal, hydraulic, and porous medium properties such as porosity, permeability, thermal conductivity, and the initial hydrate saturation. Figure 7.8 shows the sensitivity of gas pressure inside the reservoir to such properties. In each case, the concerned property is varied while the rest of the parameters are kept at the base values as assigned in Table 7.1. The results of simulation over an operation time of 60 days are presented. Figure 7.8 shows gas pressure to have a strong dependence on porosity and permeability of the reservoir.

7.4.3   Multiphase Simulation

A multiphase model with a kinetic approach for modeling CH_4 gas production from partially saturated hydrate reservoirs is presented in Sect. 7.3.1. The one-dimensional model accounts for depressurization at the production well with simultaneous CO_2 gas injection at an injection well, both located at a fixed distance from each other. Using this model, simulations have been conducted for the time evolution of CH_4 and CO_2 partial pressures, temperature, CH_4 and CO_2 hydrate and saturations, and secondary CH_4 hydrate saturations in the hydrate reservoir. CO_2 hydrates form near the injection well after some time into operation, when favorable kinetic conditions are achieved in that region. Secondary CH_4 hydrates are those that were not initially present in the reservoir but are formed during the production process, though they may dissociate again at a later point of time, under favorable kinetic conditions for dissociation. The effects of boundary conditions at production and injection wells and a sensitivity analysis of the effects of thermal, hydraulic, and

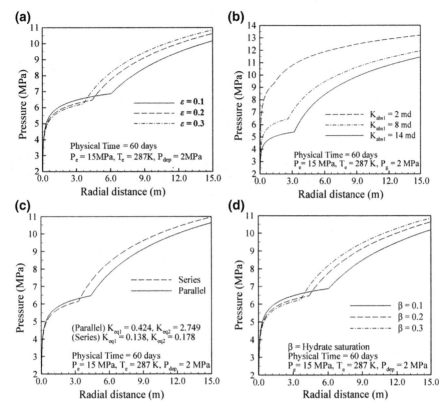

Fig. 7.8 Sensitivity of pressure distribution with respect to **a** porosity, **b** permeability, **c** thermal conductivity, and **d** initial hydrate saturation

reservoir properties on the system variables are also presented below. A major part of these studies is dedicated to the evaluation and explanation of the trends in CH_4 gas production and evolution rates under several possible parametric conditions.

7.4.4 Evolution of Pressure and Temperature Profiles

The reservoir volume analyzed in this part of the study is 10 m × 10 m × 100 m in size. The distance between the production and injection wells is taken at 100 m (10 m × 10 m being the cross-sectional area). Initially, the reservoir is partially saturated with CH_4 hydrates and contains free CH_4 gas at the equilibrium pressure and temperature of 283 K. At the beginning of production, CO_2 gas is injected at a pressure of 4.5 MPa at the injection well. At 283 K, CO_2 liquefies above ∼4.5 MPa. The chemical kinetics of hydrate formation with liquid CO_2 is not well-known. Hence, CO_2 injection pressures above 4.5 MPa were not used in the simulation. The constant pressure for depressurization at the production well is taken to be 3.5 MPa. Further details of the baseline parameter values and boundary conditions are reported in Tables 7.1 and 7.2. Figure 7.9 shows the evolution of pressure and temperature profiles over a duration of 90 days of depressurization and injection in the reservoir.

The total gas pressure, as shown in Fig. 7.9, reaches a stable value in a short span of 3 days and maintains an invariant profile for up to 30 days. However, drastic variations in partial pressures of CH_4 and CO_2 gas take place in the first 30 days of evolution. The CH_4 partial pressure variation, as shown in Fig. 7.9c, is marked by a very distinct pressure pulse which mollifies as time progresses. Such a pressure pulse has also been reported in the recent literature on CO_2 sequestration. On the other hand, the CO_2 partial pressure, as shown in the figure above, increases steadily from the injection well with time. The CO_2 gas breaks through the reservoir

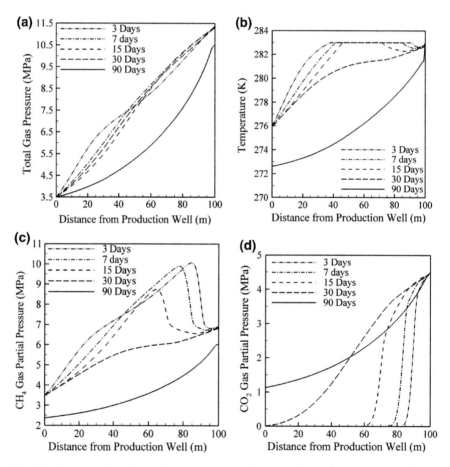

Fig. 7.9 Time evolution of **a** total gas pressure distribution, **b** reservoir temperature, **c** CH_4 gas partial pressure, **d** CO_2 gas partial pressure

(i.e., reaches the production well) in a time span of around 30 days. After 90 days of real-time simulation, the total and partial pressures of the gas components show a uniform variation across the reservoir with a much lower average value than those in simulations from 30 days of well operation. The CO_2 gas has considerable presence at the production well after 90 days.

Figure 7.9b shows the evolution of the reservoir temperature with time. During the first 15 days, temperature is lower at both the wells and higher in the middle of the reservoir. This phenomenon can be explained by its strong dependence on the thermodynamics of hydrate dissociation/formation. CH_4 gas partial pressure peaks in the middle of the reservoir and reaches values below the CH_4 hydrate equilibrium pressures near the wells during the first 15 days. This leads to the endothermic process of CH_4 hydrate dissociation near the two wells and results in a temperature drop near the well regions.

Since CH_4 pressures are much higher in the middle region of the reservoir, it can also be argued that high pressures should lead to secondary CH_4 hydrate formation. Consequently, since hydrate formation is an exothermic process, temperatures should rise above 283 K in the middle region of the reservoir. In reality, secondary CH_4 hydrate formation does take place in the middle region of the reservoir as shown in Fig. 7.10. However, temperatures at the center of the reservoir are only slightly higher ($\sim 10^{-3}$ K) than 283 K. This is because the kinetic rate constant for hydrate formation is lower than the kinetic rate constant for dissociation by $\sim 10^8$ orders in magnitude. Thus, the rate of exothermic formation is much lower than that of endothermic dissociation. Secondary CH_4 hydrate formation is important because it directly affects the local permeability and fluid porosity. Lower permeability in turn will affect gas production and flow in that region. The present estimates of secondary CH_4 hydrate saturation show it to be minute enough (of order $\sim 10^{-4}$) to cause any practical concern. However, it must be noted that kinetics of in situ hydrate formation are also not fully understood at present.

After 90 days of physical time, the temperature in the reservoir falls to lower values because of large-scale CH_4 hydrate dissociation across the reservoir. Near the production well region, it must be noted that the temperature falls to values

Fig. 7.10 Time evolution of secondary CH_4 hydrate saturation

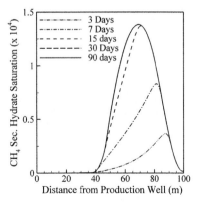

slightly below 273 K. This indicates ice formation due to the presence of water which will cause pore-plugging, thus negatively affecting the local permeability and fluid porosity. However, this interesting phenomenon points at thermal self-regulation of hydrate reservoirs. As ice will cause pore-plugging, the rate of depressurization will diminish. Thus, temperature will fall at a slower rate due to lower dissociation rates. Ice formation has not been accounted for in this model which is inaccurate if the reservoir temperature falls anywhere below 273 K.

The overall decrease in the reservoir temperature can also be corroborated by Fig. 7.11 which shows the evolution of CH_4 and CO_2 hydrate saturations and CH_4 gas production and evolution rates with respect to time. Gas production has been evaluated as the CH_4 gas mass flow rate at the production well, whereas gas evolution represents the integration of the volumetric CH_4 gas mass generation rate across the volume of the entire reservoir.

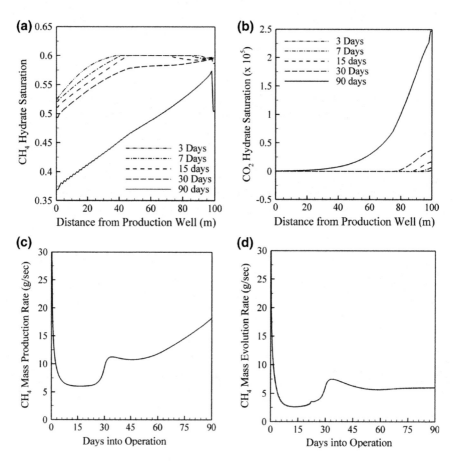

Fig. 7.11 Time evolution of **a** CH_4 hydrate saturation, **b** CO_2 hydrate saturation, **c** CH_4 mass production rate, and **d** CH_4 mass evolution rate

Figure 7.11a shows that the decomposition of CH_4 hydrate is greater near the well regions due to lower CH_4 pressures, as explained earlier in the context of temperature. It is observed that there is a good degree of similarity between the CH_4 hydrate saturation and temperature profiles (as shown in Fig. 7.11b) for the first 30 days. Figure 7.11b shows a gradual increase in CO_2 hydrate formation over time, as expected, with CO_2 gas entering deeper into the reservoir. Analogous to secondary methane hydrates, it must be noted that CO_2 hydrate formation occurs at very low rates as compared to dissociation owing to the large difference in the kinetic constants for the two processes.

Figure 7.11c shows variation of mass production rate of CH_4 gas at the depressurization well with respect to time. Production from the reservoir can be visualized in the following manner. At the very initial time, when the CH_4 hydrates in the reservoir are at equilibrium conditions, depressurization extracts the free gas available in the near-well zone of the reservoir thus showing very high production rates. This leads to lowering of pressure in the near-well zone which subsequently initiates hydrate dissociation, thus leading to further release of CH_4 gas. Finally after some time, both flow and hydrate dissociation substantially contribute to the total production.

Figure 7.11d shows the time variation of CH_4 gas mass evolution rate (integrated over the entire reservoir). Evolution rate in the reservoir is initially high, and it gradually decreases to a constant value over around 9–18 days. The evolution rate then starts to climb again and is marked by a sharp change in between 21 and 23 days. Figure 7.12 can be used to understand the sharp change in the evolution rate during this period. Figure 7.12a presents a magnified view of the time variation of CH_4 evolution rate for the time duration between 15 and 25 days. The evolution rate at any given time is simply an integration of the volumetric mass generation rate of CH_4 gas in the reservoir as shown in Fig. 7.12b (for 15, 21, 23, and 25 days). Figure 7.12c shows the CH_4 gas partial pressure profile, and Fig. 7.12d shows the CH_4 hydrate equilibrium pressure profile, providing support to the trends in Fig. 7.12a–b.

The sharp change in evolution rate can now be understood in the following manner. The CH_4 hydrate equilibrium pressure is a function of temperature and is marked by sharp changes in its profile at 15 and 21 days. Such sharp changes are, however, absent in its profile at 23 and 25 days. Now, the volumetric mass generation rate of CH_4 gas (at any given time and location in the reservoir) is a linear function of the difference between the CH_4 gas partial pressure and CH_4 hydrate equilibrium pressure. Figure 7.12b clearly depicts the sharp peaks in the volumetric mass generation rate at 15 and 21 days, which are coincidently absent for 23 and 25 days. Thus, integration of this volumetric mass rate over the volume of the entire reservoir results in a sharp change in the evolution rate between 21 and 23 days.

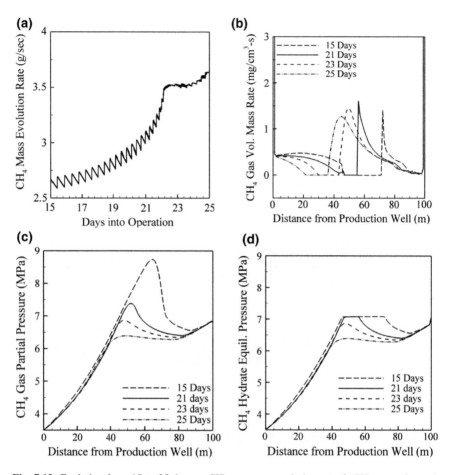

Fig. 7.12 Evolution from 15 to 25 days. **a** CH_4 gas mass evolution rate, **b** CH_4 gas volumetric mass production rate, **c** CH_4 gas partial pressure, and **d** CH_4 hydrate equilibrium pressure

7.4.5   Sensitivity Analysis: Thermal Conductivity

Both CH_4 production and CO_2 sequestration could be strongly sensitive to the hydraulic and thermophysical properties of the porous media. Due to brevity, sensitivity analysis with respect to only two parameters will be presented here: thermal conductivity and reservoir porosity.

Equivalent thermal conductivity of the composite medium can be calculated for parallel and series modes of energy transport (Eq. 7.30). An average of the equivalent conductivities from these two modes has also been considered in the analysis. The present discussion on sensitivity analysis is not concerned with the uncertainties in the absolute conductivity values for each of the constituents of the system. Instead, the spatial orientation of conductivity that primarily determines heat transfer is of interest.

Figure 7.13 shows the effect of the mode of conduction in the reservoir on the system variables. The parallel model gives a higher equivalent thermal conductivity than the series. Higher equivalent thermal conductivity in the reservoir leads to greater heat transfer from the higher (constant) temperature injection well boundary into the reservoir, eventually resulting in a more uniform temperature distribution in that region, Fig. 7.13b. Near the production well, the effect of endothermic hydrate dissociation on local temperatures is quite strong for thermal conductivity to have a measureable influence. Pressure profiles and CH_4 hydrate saturations, as seen in Fig. 7.13a and c, respectively, remain mostly unaffected by the choice of the thermal conductivity model. Figure 7.13d corroborates the fact that lower temperatures favor CO_2 hydrate formation near the injection well.

Figure 7.14 shows that the choice of thermal conductivity model has a small effect on CH_4 mass production and evolution rates and secondary CH_4 hydrate formation. Higher thermal conductivity results in higher production rates and eventually higher evolution rates (after ~ 28 days).

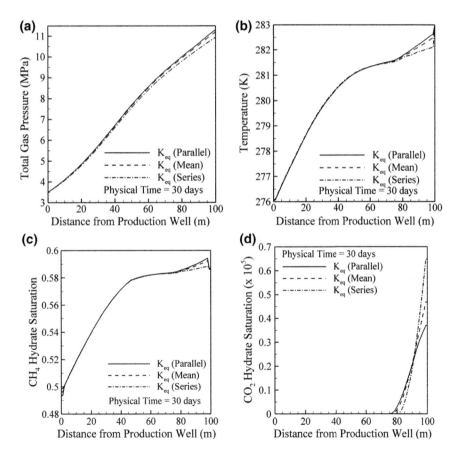

Fig. 7.13 Variation with respect to the choice of thermal conductivity for: **a** total gas pressure, **b** reservoir temperature, **c** CH_4 hydrate saturation, **d** CO_2 hydrate saturation

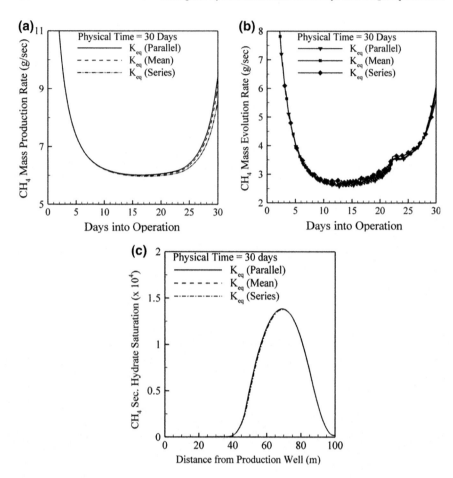

Fig. 7.14 Variation with respect to the mode of heat conduction for: **a** CH_4 gas mass production rate, **b** CH_4 gas mass evolution rate, **c** secondary CH_4 hydrate saturation

7.4.6 Sensitivity Analysis: Medium Porosity

For this analysis, medium porosity values of 0.18, 0.28, and 0.38 were considered for a reservoir made of Berea Stone. In the present study, absolute permeability is considered to be a function of the effective fluid porosity and therefore and independent sensitivity analysis with permeability was not considered necessary. Figure 7.15 shows the effect of medium porosity on the system variables in the reservoir.

Porosity of the gas hydrate bearing medium has a very strong effect on each of the physical variables including pressure, temperature, and phase saturation. Higher porosity has basically two major implications: (1) higher absolute permeability and (2) higher volume of CH_4 hydrate at a given saturation. As seen in Fig. 7.15a, the total gas pressure varies drastically with respect to change in medium porosity.

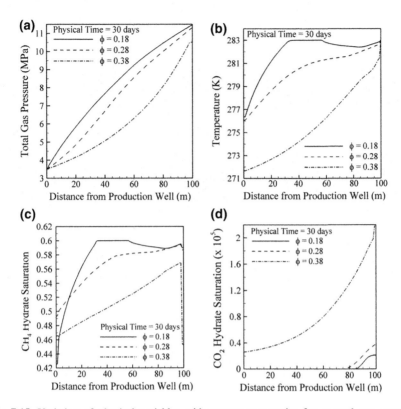

Fig. 7.15 Variation of physical variables with respect to porosity for: **a** total gas pressure, **b** reservoir temperature, **c** CH_4 hydrate saturation, **d** CO_2 hydrate saturation

For higher porosity, flow rates inside the reservoir increase and the resulting rate of depressurization is much higher. This leads to much lower reservoir pressures. Similarly, the reservoir temperature is strongly influenced by medium porosity, as shown in Fig. 7.15b. Higher rate of depressurization due to higher porosity leads to additional dissociation of CH_4 hydrates. This endothermic process contributes substantially to the lowering of temperatures across the reservoir.

Remarkably, for a high porosity value of 0.38, the temperatures near the production well fall below 273 K, indicating that ice formation will take place. Ice formation will reduce the effective fluid porosity thus reducing the permeability of the medium. This, in turn, will lead to a reduction in gas flow rates. The phenomenon yet again points to the possibility of thermal self-regulation of hydrate reservoirs. Further, it indicates that in the long term, higher porosities do not necessarily guarantee higher production rates.

Figure 7.15c points toward rapid CH_4 hydrate dissociation in the reservoir for high porosities. However, for a low porosity value of 0.18, the CH_4 hydrate saturation is lower near the production well when compared to saturation in case of higher porosities. This occurrence can be explained as a result of the fact that lower porosity and permeability in the reservoir impedes flow of CH_4 gas from the far end of the reservoir, being already blocked by a pressure pulse in the CH_4 partial pressure profile. For a given depressurization pressure, even more rapid depressurization of the reservoir takes place in the near-well region for a low porosity medium. This leads to additional CH_4 hydrate dissociation. It must, however, be noted that this rate of depressurization is greater than the rates in case of higher porosities only near the production well. When considered over the length of the reservoir, overall depressurization is greater when porosity is higher.

Figure 7.15d shows that CO_2 hydrate saturation is higher in a high porosity medium. Hydrate formation is strongly dependent on surface area available for nucleation. In a higher porosity medium, this area is larger, thus supporting hydrate formation. Moreover, due to higher porosity, infiltration of CO_2 gas in the medium is faster, assisting formation of CO_2 hydrates.

Figure 7.16 shows the effect of medium porosity on CH_4 gas production, evolution rates, and secondary CH_4 hydrate formation. As expected from the pressure and hydrate saturation profiles in Fig. 7.15, high porosity of a medium will provide

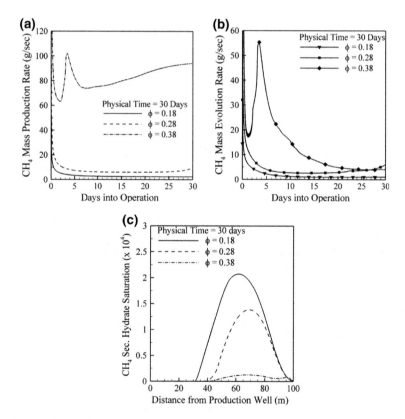

Fig. 7.16 Variation of gas production parameters with respect to porosity for: **a** CH_4 gas mass production rate, **b** CH_4 gas mass evolution rate, and **c** secondary CH_4 hydrate saturation

with much higher production and evolution rates due to rapid depressurization. The secondary CH_4 hydrate formation is favored in a high porosity medium in a manner similar to which CO_2 hydrate formation is favored.

7.5   Closure

A large amount of methane stays in hydrate form all across the coastal region. Recovery of methane from hydrates may provide the key to the global energy security of future generations. Methane combustion, however, produces CO_2, the leading agent for the greenhouse effect. Sequestering anthropogenic CO_2 in hydrate reservoir may enhance methane production while cleaning the atmosphere from the combustion products of methane. Development of successful technologies of CH_4 production and CO_2 sequestration requires investigation on fundamental aspects of transport in porous media. Transport phenomena, in such cases, are modeled around coupled, multiscale–multiphysics formulations that are highly nonlinear in nature. Development of large-scale, efficient numerical simulators, as well as efficient laboratory-scale experiments for the model validation, will continue to challenge the porous media research community in coming decades.

References

1. E.D. Sloan, in *Clathrate Hydrates of Natural Gases*, 2nd ed. Chemical Industries, vol 73, (Marcel Dekker, New York, 1998), p. 705
2. E.D. Sloan, C.A. Koh, in *Clathrate Hydrates of Natural Gases*, 3rd ed. Chemical Industries, (Taylor & Francis CRC Press: Boca Raton, FL, 2007)
3. K.A. Birkedal, Hydrate Formation and CH_4 Production from Natural Gas Hydrates. A Master's Thesis in Reservoir Physics, University of Bergen, Norway, 2009
4. E.D. Sloan, in *Hydrate Engineering. SPE Monograph Series*, vol 21, (Society of Petroleum Engineers, Richardson, 2000)
5. K.A. Kvenvolden, Gas hydrates—geological perspective and global change. Rev. Geophys. **31**(2), 172–187 (1993)
6. R.D. McIver, Role of naturally occurring gas hydrates in sediment transport. AAPG Bull. **66**, 789–792 (1982)
7. T.S. Collett, S.R. Dallimore, in *Detailed analysis of gas hydrate induced drilling and production hazards*. Proceedings of the 4th International Conference on Gas Hydrates, (Yokohama, Japan, 2002)
8. G.T. MacDonald, Role of methane clathrates in past and future climates. Clim. Change **16**, 247–281 (1990)
9. N. Goel, In situ methane hydrate dissociation with carbon dioxide sequestration: Current knowledge and issues. J. Petrol. Sci. Eng. **51**, 169–184 (2006)
10. K. Hester, P.G. Brewer, Clathrate hydrates in nature. Ann. Rev. Mar. Sci. **1**, 303–327 (2009)
11. A.A. Trofimuk, N.V. Cherskiy, V.P. Tsarev, Accumulation of natural gases in zones of hydrate—formation in the hydrosphere. Dokl. Akad. Nauk SSSR **212**, 931–934 (1973)
12. V.A. Soloviev, Global estimation of gas content in submarine gas hydrate accumulations. Russ. Geol. Geophys. **43**, 609–624 (2002)

13. M.D. Max, A.H. Johnson, W.P. Dillon, in *Economic Geology of Natural Gas Hydrate*, vol 9, 2006
14. B.S. Pierce, T.S. Collett, in *Energy Resource Potential of Natural Gas Hydrates*. 5th Conference & Exposition on Petroleum Geophysics, (Hyderabad, India, 2004), pp. 899–903
15. K.A. Kvenvolden, in *A Primer on Gas Hydrate*, ed. by D.G. Howell et al. The Future of Energy Gases, US Geological Survey Professional Paper 1570, (1993), pp. 279–291
16. E.D. Sloan, Fundamental principles and applications of natural gas hydrates. Nature **426**, 353–363 (2003)
17. K. Ohgaki, K. Takano, H. Sangawa, T. Matsubara, S. Nakano, Methane exploitation by carbon dioxide from gas hydrates-phase equilibria for CO$_2$–CH$_4$ mixed hydrate system. J. Chem. Eng. Japan **29**(3), 478–483 (1996)
18. D. Sloan, C.A. Koh, *Clathrate Hydrates of Natural Gases*, 3rd ed. (CRC Press, Boca Raton, Fl, 2008)
19. G. Ersland, J. Husebø, A. Graue, B. Kvamme, B.A. Baldwin, J.J. Howard, J. Stevens, in *Measurements of Gas Permeability and Non-Darcy Flow in Gas-Water-Hydrate Systems*. 6th International Conference on Gas Hydrates, (Vancouver, BC, Canada, July 6–10, 2008)
20. H. Lee, Y. Seo, Y.-T. Seo, I.L. Moudrkovski, J.A. Ripmeester, Recovering methane from solid methane hydrate with carbon dioxide. Angew. Chem. **115**(41), 5202–5205 (2003)
21. H.A. Phale, T. Zhu, M.D. White, B.P. McGrail, *Simulation study on injection of CO$_2$-Microemulsion for Methane Recovery From Gas-Hydrate Reservoirs* (SPE Gas Technology Symposium, Calgary, Alberta, Canada, 2006)
22. H.C. Kim, P.R. Bishnoi, R.A. Heideman, S.S.H. Rizvi, Kinetics of methane hydrate decomposition. Chem. Eng. Sci. **42**(7), 1645–1653 (1987)
23. M.B. Kowalsky, G.J. Moridis, Comparison of kinetic and equilibrium reaction models in simulating gas hydrate behavior in porous media. Energy Convers. Manag. **48**, 1850–1863 (2007)
24. Y.F. Makogon, Hydrates of Natural Gas. Translated form Russian by Cieslesicz, WJ, Penn Well, Tulsa, OK, 1974
25. N.N. Verigin, I.L. Khabibullin, G.A. Khalikov, Izv. Akad. Nauk. SSSR, Mekhanika Zhidkosti Gaza, **1**, 174 (1980)
26. D. Lysne, in *Hydrate plug dissociation by pressure reduction*, ed. by E.D. Sloan Jr., J. Happel, M.A. Hnatow. International Conference on Natural Gas Hydrates, vol 715, (Academy of New York, Academy of Sciences, 1993), pp. 714–717
27. C. Ji, G. Ahmadi, D.H. Smith, Natural gas production from hydrate decomposition by depressurization. Chem. Eng. Sci. **56**, 5801–5814 (2001)
28. G. Ahmadi, C. Ji, D.H. Smith, Numerical solution for natural gas production from methane hydrate dissociation. J. Petrol. Sci. Eng. **41**, 269–285 (2004)
29. G. Ahmadi, C. Ji, D.H. Smith, Natural gas production from hydrate dissociation: An axisymmetric model. J. Pet. Sci. Eng. **58**, 245–258 (2007)
30. S. Gerami, M. Pooladi-Darvish, Predicting gas generation by depressurization of gas hydrates where the sharp-interface assumption is not valid. J. Petrol. Sci. Eng. **56**, 146–164 (2007)
31. M.H. Yousif, H.H. Abass, M.S. Selim, E.D. Sloan, in *Experimental and Theoretical Investigation of Methane-Gas-Hydrate Dissociation in Porous Media*. SPE paper 18320, 1991
32. X. Sun, N. Nanchary, K.K. Mohanty, 1-D modeling of hydrate depressurization in porous media. Transp. Porous Media **58**, 315–338 (2005)
33. H. Hong, M. Pooladi-Darvish, Simulation of depressurization for gas production from gas hydrate reservoir. University of Calagy, J. Canadian Pet. Technol. **44**, 39–46 (2005)
34. G.J. Moridis, in *Numerical Studies of Gas Production from Methane Hydrates*. SPE Paper 75691 presented in the SPE Gas Technology Symposium held in Calgary, Alberta, 2002
35. M.D. White, B.P. McGrail, in STOMP-HYD: a new numerical simulator for analysis of methane hydrate production from geologic formations. Proceedings of 2nd International Symposium on Gas Hydrate Technology, 1–2 Nov 2006

36. A. Khetan, M.K. Das, K. Muralidhar, Analysis of methane production from a porous reservoir via simultaneous depressurization and CO_2 sequestration. Spec. Top. Rev. Porous Media **4**(3), 237–252 (2013)
37. K. Muralidhar, M.K. Das, hydrate reservoir—methane recovery and CO_2 disposal. Proceedings of Indian National Science Academy **81**(4), 787–800 (2015)
38. G.J. Moridis, M.T. Reagan, in *Strategies for Gas Production From Oceanic Class 3 Hydrate Accumulations*. Offshore Technology Conference-18865, Houston, Texas, 30 April–3 May 2007
39. G.J. Moridis, M.T. Reagan, in *Gas Production From Class 2 Hydrate Accumulations in the Permafrost*. SPE 110858, Annual Technical Conference and Exhibition, Anaheim, California, USA, 11–14 Nov 2007
40. R. Boswell, Resource potential of methane hydrate coming into focus. JPSE **56**, 9–13 (2007)
41. Y.T. Seo, H. Lee, Multiple-phase hydrate equilibria of the ternary carbon dioxide, methane, and water mixtures. J. Phys. Chem. B **105**(41), 10084–10090 (2001)
42. Y.T. Seo, H. Lee, J.H. Yoon, Hydrate phase equilibria of the carbon dioxide, methane, and water system. J. Chem. Eng. Data **46**(2), 381–384 (2001)
43. S. Hirohama, Y. Shimoyama, A. Wakabayashi, S. Tatsuta, N. Nishida, Conversion of CH_4-hydrate to CO_2-hydrate in liquid CO_2. J. Chem. Eng. Japan **29**(6), 1014–1020 (1996)
44. M.B. Clennell, M. Hovland, J.S. Booth, P. Henry, W.J. Winters, Formation of natural gas hydrates in marine sediments 1. conceptual model of gas hydrate growth conditioned by host sediment properties. J. Geophys. Res. **104**(B10), 22985–23003 (1999)
45. W. Rice, Proposed system for hydrogen production from methane hydrate with sequestering of carbon dioxide hydrate. J. Energy Resour. Technol. (ASME) **125**(4), 253–256 (2003)
46. W. Rice, Hydrogen production from methane hydrate with sequestering of carbon dioxide. Int. J. Hydrogen Energy **31**(14), 1955–1963 (2006)
47. H.A. Phale, T. Zhu, M.D. White, B.P. McGrail, *Simulation Study on Injection of CO_2-Microemulsion for Methane Recovery from Gas-Hydrate Reservoirs*, University of Alaska Fairbanks and Pacific Northwest Natl. Laboratory, SPE Paper-100541, 2006
48. M. Uddin, D.A. Coombe, D. Law, W.D. Gunter, Numerical studies of gas hydrate formation and decomposition in a geological reservoir. Contributed by the Petroleum Division of ASME for publication in the J. Energy Resour. Technol., **130**, 032501 (2008)
49. CMG STARS®, in *Advanced Process and Thermal Reservoir Simulator*. (Computer Modelling Group Ltd., Calgary, Alberta, Canada)
50. S. Adisasmito, R.J. Frank, E.D. Sloan, Hydrates of carbon dioxide and methane mixtures. J. Chem. Eng. Data **36**, 68–71 (1991)
51. B.J. Anderson, M.Z. Bazant, J.W. Tester, and B.L., Trout, Application of the cell potential method to predict phase equilibria of multicomponent gas hydrate systems. J. Phys. Chem. B, **109**, 8153–8163 (2005)
52. R.C. Reid, J.M. Prausnitz, B.E. Poling, *The Properties of Gases & Liquids* (McGraw-Hill Book Company, New York, 1987)
53. M.S. Selim, E.D. Sloan, Heat and mass transfer during the dissociation of hydrates in porous media. AIChE J. **35**, 1049–1052 (1989)
54. G.K. Anderson, Enthalpy of dissociation and hydration number of carbon dioxide hydrate from the Clapeyron equation. J. Chem. Thermodyn. **35**, 1171–1183 (2003)
55. G.K. Anderson, Enthalpy of dissociation and hydration number of methane hydrate from the clapeyron equation. J. Chem. Thermodyn. **36**, 1119–1127 (2004)

Chapter 8
Closure

Malay K. Das

The book is an attempt toward connecting time-honored as well as contemporary theories of porous media with rapidly growing engineering applications. In recent decades, applications of transport in porous media have grown to such an extent that no book or monograph can claim exhaustive coverage of all relevant topics. In this context, the present work may be treated as a starting point for a conversation between the classical and contemporary theories, followed by a contextualization of theories useful for engineering applications. The relevance of the present book, therefore, lies in exploring the rigor of theory of transport in porous media and its utility for real-life analysis.

The book introduces phenomenological theories of transport in porous media and contemporary mesoscopic models. While these two approaches together cover many traditional and emerging processes and devices, use of the mesoscopic tools requires reconstruction of the pore-scale geometry using advanced imaging and tomography techniques. The subject of reconstructing the pore-scale geometry is inherently complex and calls for a separate monograph. On the other hand, transport in porous media may also be modeled at a continuum scale without volume-averaging the governing equations. Such techniques, known as the direct numerical simulation, involve idealization of the pore-scale geometry followed by the solution of classical conservation equations within the pore space for each of the fluid and solid phases. Such an approach is again vast and has not been included in the present volume. Finally, the present book discusses the classical and mesoscopic theories from an application viewpoint desisting rigorous derivation of the intermediate steps. The present book, therefore, acknowledges the multiplicity of treatments in porous media and paves the way for future work on connecting newer theories with relevant applications. A few additional subjects that require specialized treatment and discussion are listed below.

i. Evaporation and condensation in porous media with complete, non-empirical treatment of capillarity;
ii. Multiphase flow in porous media with hydrophobic surfaces;

© Springer International Publishing AG 2018
M.K. Das et al., *Modeling Transport Phenomena in Porous Media
with Applications*, Mechanical Engineering Series,
https://doi.org/ 10.1007/978-3-319-69866-3_8

iii. Membrane models of transport derived from theories of porous media;
iv. Transport mechanisms in hierarchical systems;
v. A framework of measurement techniques for porous media.

The primary purpose of the present book is to connect theories of porous media with important engineering applications. A large number of such applications are directed toward energy and environment-related technologies. The energy-environment context is highlighted in this book, through the discussions on batteries, fuel cells, methane recovery, and carbon dioxide sequestration. Similarly, recent decades have also seen the drive for innovative energy-efficient devices. Heat exchanger forms an integral part in many such energy-conversion devices. The book, therefore, includes a chapter on novel approaches in heat exchanger analysis and design. The concise volume, nonetheless, omits several important applications. One particularly crucial topic is the high-temperature applications in porous media. Investigation of reacting flows including combustion and high-temperature fuel cells falls in this category. Inclusion of such applications, in a book, further requires a theoretical background of radiation and reactions in porous media. While acknowledging the importance of high-temperature applications, the authors set the task aside for a separate stand-alone volume.

Overall, the present book does not intend to include all aspects of porous media transport. Instead, the book precisely serves the purpose of setting the baseline for porous media research and laying out the necessary tools to burgeoning researchers and graduate students joining the field. Its highlight, therefore, lies in the variety of topics covered and the moderation of the depth at which the topics are introduced.

Index

© Springer International Publishing AG 2018
M.K. Das et al., *Modeling Transport Phenomena in Porous Media with Applications*, Mechanical Engineering Series,
https://doi.org/ 10.1007/978-3-319-69866-3

Printed by Printforce, the Netherlands